IGNORING THE APOCALYPSE

Recent Titles in Politics and the Environment

Gaia's Revenge: Climate Change and Humanity's Loss
P. H. Liotta and Allan W. Shearer

IGNORING
THE APOCALYPSE

Why Planning to Prevent
Environmental Catastrophe Goes Astray

DAVID HOWARD DAVIS

Politics and the Environment
P. H. Liotta, Series Editor

Westport, Connecticut
London

Library of Congress Cataloging-in-Publication Data

Davis, David Howard.
 Ignoring the apocalypse : why planning to prevent environmental
catastrophe goes astray / by David Howard Davis.
   p. cm. — (Politics and the environment, ISSN: 1932–3484)
 Includes bibliographical references and index.
 ISBN-13: 978–0–275–99663–5 (alk. paper)
 ISBN-10: 0–275–99663–8 (alk. paper)
 1. Environmental degradation—Prevention. 2. Global warming—
Prevention. 3. Climatic changes—Prevention. 4. Nuclear warfare—
Prevention. 5. Overpopulation—Prevention. 6. Energy conservation.
I. Title.
 GE140.D382 2007
 363.7—dc22   2007008640

British Library Cataloguing in Publication Data is available.

Library of Congress Catalog Card Number: 2007008640

ISBN-13: 978–0–275–99663–5
ISBN-10: 0–275–99663–8
ISSN: 1932–3484

First published in 2007

Praeger Publishers, 88 Post Road West, Westport, CT 06881
An imprint of Greenwood Publishing Group, Inc.
www.praeger.com

Printed in the United States of America

The paper used in this book complies with the
Permanent Paper Standard issued by the National
Information Standards Organization (Z39.48–1984).

10 9 8 7 6 5 4 3 2 1

This book is dedicated to
Thomas R. Lopez, PhD

Contents

Series Foreword

The key focus of the Praeger Politics and Environment series is to explore the interstices between environmental, political, and security impacts in the twenty-first century. To those intimately involved with these issues, their immediacy and importance are obvious. What is not obvious to many, nonetheless—including those involved in making decisions that affect our collective future—is how these three critical issues are in constant conflict and frequently clash. Yet today, more than ever before in history, intersecting environmental, political, and security issues have an impact on our lives and the lives of those who are to come.

In examining the complex interdependence of these three impact effects, the study of environmental and security issues should recognize several distinct and pragmatic truths: First, international organizations today are established for and focus best on security issues; thus while it remains difficult for these organizations to address environmental threats, challenges, and vulnerabilities, it makes imminently better sense to reform what we have rather than constantly invent the new organization that may be no better equipped to handle current and future challenges. Second, the new protocols must continue to be created, worked into signature, and managed under the leadership of states through international organizations and cooperative regimes. Third, and incorporating the reality of these previous two truths, we should honestly recognize that environmental challenges can best be presented in terms that relate to security issues. To that end, it is sensible to depict environmental challenges in language that is understandable to decision makers most familiar with security impacts and issues.

There is benefit and danger in this approach, of course. Not all security issues involve direct threats; some security issues, as with some political processes, are far more nuanced, more subtle, and less clearly evident. I would argue further—as I have been arguing for several decades now—that it remains a tragic mistake to couch all security issues in terms of threat. To the contrary, what I term "creeping vulnerabilities"—population growth, disease, climate change, scarcity of water and other natural resources, decline in food production, access, and availability, soil erosion and desertification, urbanization and pollution, and the lack of effective warning systems—can come to have a far more devastating impact if such issues are ignored and left unchecked over time. In the worst possible outcomes, vulnerabilities left unchecked over time *will* manifest themselves as threats.

In its most direct, effective, and encompassing assessment, environmental security centers on a focus that seeks the best response to changing environmental conditions that have the potential to reduce stability and affect peaceful relationships and—if left unchecked—could lead to the outbreak of conflict. This working definition, therefore, represents the vital core of the Praeger Politics and the Environment series.

Environmental security emphasizes the sustained viability of the ecosystem, while recognizing that the ecosystem itself is perhaps the ultimate weapon of mass destruction. In 1556 in Shensi province, for example, tectonic plates shifted, and by the time they settled back into place, 800,000 Chinese were dead. Roughly 73,500 years ago, a volcanic eruption in what is today Sumatra was so violent that ash circled the earth for several years, photosynthesis essentially stopped, and the precursors to what is today the human race amounted to only several thousand survivors worldwide. The earth itself, there can be little doubt, is the ultimate weapon of mass destruction. Yet from an alternate point of view, mankind itself is the ultimate threat to the earth and the earth's ecosystem.

Three decades ago, the environmentalist Norman Myers wrote that national security is about far more than fighting forces and weaponry. National security must also include issues of environment and environmental impact—from watersheds to climate impact—and these factors must figure in the minds of military experts and political leaders. Myers's words today remain as prophetic, and deadly accurate, as ever.

In this volume of the Politics and Environment series, David Howard Davis's *Ignoring the Apocalypse: Why Planning to Prevent Environmental*

Catastrophe Goes Astray centers on four "apocalyptic" threats that face us: the increase in greenhouse gases leading to global warming, the presence and growing likelihood in use of nuclear weapons, significant population and demographic shifts, and an energy crisis that is no longer looming but is now well underway.

Davis asks whether strategic planning, in the United States in particular, is different for dangers that are truly apocalyptic—ones that could end life on the planet or at least modern economic prosperity. Moreover, he illustrates how policy makers tend to ignore challenges such as oil shortages, nuclear war, and environmental change—despite the warning signs—and, when planning does take place, shows how these plans often go astray.

Environmentalists often predict an apocalypse is coming: The earth will heat up like a greenhouse. We will run out of energy. Overpopulation will lead to starvation and war. Nuclear winter will devastate all organic life. We have, of course, grown desensitized to many such prophecies of doom. Davis argues, nonetheless, that the time and the need for strategic disaster planning are more pressing than ever.

In *Ignoring the Apocalypse,* Davis shows that we need to be attentive to the environmental challenges before us. If we do not heed the warning signs, then we imperil ourselves and our future. As Davis notes, "An apocalypse predicts the end of an era." In more ways than one, we have already entered a new era—one where we can no longer afford to be oblivious.

P. H. Liotta
Executive Director
Pell Center for International Relations and Public Policy
Newport, Rhode Island

Preface

During the 30 years I have been working in the environmental arena, I have pondered that some threats are greater than others. Some will actually end life on planet Earth, or at least end prosperous and democratic society. Greenhouse gases could cause the globe to warm up, destroying our agriculture so that we will not have food. A nuclear war with as few as two hundred bombs could generate so much smoke and dust that the summer weather would be like winter, too cold for plants to grow. Although overpopulation may seem to be the opposite of the extinction of the human species, it could lead to poverty, starvation, and border wars over farmland. The energy crisis also could reduce us to poverty and ignite wars; indeed, some say that these wars have already begun. I have labeled these four threats apocalyptic and thought it would be interesting to write a book solely about them.

Americans like to frame issues using the image of the biblical Apocalypse. Dating back to colonial New England, we have called for people to end their foolish ways and turn back to God or else he will destroy the world. Environmentalists have updated the message. Rachel Carson wrote that if we continued to spread pesticides, all the birds would die and none would be able to sing in the spring. Others have predicted an end to nature. Still others have foretold that the greenhouse effect would be apocalyptic. While few environmentalists are overtly religious, they often use this image.

As a nation we Americans have been concerned about these apocalyptic dangers, sometimes dating back for a century. For example, in 1908 President Theodore Roosevelt convened the White House Conference on Natural Resources, resulting in a warning that we faced a shortage of oil and other minerals. Oil shortages were predicted again in 1952 and 1956. With the advent of mainframe computers, scientists became able to make specific forecasts based on statistics. In 1972 the Club of Rome predicted that the entire world economy would "crash" in about the year 2020. Yet in spite of 65 years of projections, the 1973 oil crisis came as a surprise. Time after time we have ignored threats, even the most catastrophic ones.

As a nation we sometimes plan for a threat, and sometimes do nothing at all. Once the oil embargo began, presidents Nixon and Carter and Congress made many plans and passed many laws. However, the national energy plan did not actually help, and it withered away. Once it became obvious that carbon dioxide and other gases were causing warming, the United States and other countries developed a plan, embodied in the Kyoto Protocol, to control the problem. However, a new president, George W. Bush, decided that we would not sign the treaty. Once the risk of nuclear winter was understood in 1983, we did not even bother planning.

The book takes an American perspective, asking how the United States, in conjunction with our friends and allies, tries to confront potential catastrophes. It starts by discussing how environmentalists have adopted the apocalyptic rhetoric of the Bible and goes on to consider the popularity of the genre to Americans. It would be hard to find a more exciting image than the end of the world as described by John in his Revelation or by Daniel in his prophecy.

Modern predictions depend on scientists, statisticians, and econometricians, not ancient prophets. The first step is to believe that the future will be different from the present, a belief not accepted until the modern era. Sir Thomas More offered his vision in *Utopia* in 1516. Our founders like Franklin and Jefferson planned a new republic that would foster a new kind of citizen. The next step is to understand the science of the situation, hence geologists study potential oil fields, demographers analyze population trends, and physicists invent the atom bomb. Finally we need to get some real numbers about a problem. Statisticians examined the data and, since about 1970, have been able to calculate probable outcomes. The Club of Rome report was the pioneering effort, the first to use massive sets of statistics on mainframe computers. This was the brainchild of one man—Aurelio Peccei—a visionary Italian businessman who had been pondering the Predicament of Mankind for many years. It typifies planning as prediction.

The next four chapters look at apocalyptic threats. Chapter three examines the energy crisis, which arrived unexpectedly on October 17, 1973 (in spite of many warning over the years). The Nixon administration's response was to pull out wartime plans based on command and control. These dated to the first and second World Wars, as well as to the New Deal. This was often called central planning, meaning planning from Washington, and was embodied in the national energy plan of President Jimmy Carter. The strategy was totally rejected by Ronald Reagan, who favored the free market. Although the market largely solved the problem, presidents and Congress cannot let go entirely of central planning. Even George W. Bush issued a national energy plan, created by Vice President Cheney.

Chapter four focuses on overpopulation. The risk of there being too many people in the world has been apparent since the 1950s. In 1965 President Johnson warned Congress of "the population explosion." At present the United States does not have an explicit population policy at home. Planning is absent. Yet since colonial days, we have been concerned with immigration. In the nineteenth century we welcomed people coming from Europe (with a few exceptions). In 1924 our policy shifted to stabilizing our population according to the existing ethnic mix of English, Irish, Germans, Italians, and so forth. Policy changed in 1965 to welcome immigrants from everywhere, and in 1986 to controlling the illegals and those from Mexico. We do have an explicit population policy abroad, but this has shifted from one administration to another. The crux is abortion and birth control. The one-child policy of China has been at the center of many debates. A final issue is eugenics, which shaped state laws during the first half of the twentieth century, but dropped out of discussion after the Nazi atrocities. Yet today our physicians perform 100,000 amniocentesis tests annually, and President Bush and Congress are clashing on stem cell research and prenatal testing.

Chapter five examines the environmental threat of nuclear war. In an all-out apocalypse, 50 to 100 million people would die of the blast and radiation. In the following weeks dust and soot would obscure the rays of the sun, so plants would die, then the animals that ate the plants, and finally humans as our supplies dwindled. The sunlight would only be as strong as on a winter day. After Hiroshima, the first nine years of the atomic arms race had no planning at all. We simply manufactured bombs until we had 2,000. At this point, key leaders recognized the danger, but it took another seven years until we were able to negotiate the first of about 10 treaties with the Soviet Union. These do not assure safety, but do decrease the risk a little. The demise of the Soviet Union has lessened the danger, yet many old warheads are not accounted for. Rogue countries like North Korea and Iran are likely to possess warheads on missiles within a few years. A difficulty for political discussion is that most of the detailed scientific knowledge about nuclear weapons is secret.

Finally, chapter six analyzes planning to control global warming. The scientific phenomenon was discovered over a century ago, but was ignored until 1985. Even today many government and business leaders continue to deny the danger. Climate change is a subject that cannot be understood without scientific training. Gathering information requires international cooperation, and taking practical steps does too. We have a model to emulate in the Montreal Protocol to protect the ozone layer and have copied it for the Kyoto Protocol. While the United States cooperated with other countries on ozone, we have not done so on greenhouse gases. President Bush says that the Kyoto Protocol would hurt our economy and is not fair because it exempts China and India.

My book has evolved over six years, during which time I have had the opportunity to travel to Europe, China, Peru, and Australia, as well as around the United States, to talk to experts and observe the situation. Pieces of it come from my experiences teaching, writing other books, and working for the U.S. Environmental Protection Agency, the Department of the Interior, the Congressional Research Service, and as a business consultant. In my writing I have enjoyed the help of many colleagues who have read chapters and advised me. At the University of Toledo these are Larry Connin of the Honors Program, Michael Phillips and Donald Stierman of Environmental Science, Craig Hatfield of Geology, Constantine Theodosiou of Physics, Michael Jakobson of History, and David Wilson of Political Science. Lynn McCallum of St. Andrew's Church helped as well. I also appreciate critiques by Phyllis Piotrow of the Johns Hopkins School of Public Health, by Ralph Menning of Kent State University, and by Bruce Ogilvie.

Abbreviations

This list does not include common abbreviations like EPA, NATO, NASA, and USSR.

ABM	Anti-ballistic missile
AEC	Atomic Energy Commission
AID	Agency for International Development
CFC	Chlorofluorocarbons
EEI	Edison Electric Institute
FCCC	Framework Convention on Climate Change
FEA	Federal Energy Administration
IAEA	International Atomic Energy Agency
ICBM	Intercontinental ballistic missile
IEES	International Energy Evaluation System
IMF	International Monetary Fund
INF	Intermediate-Range Nuclear Forces
IPCC	Intergovernmental Panel on Climate Change
NIRA	National Industrial Recovery Act of 1933
NOAA	National Oceanic and Atmospheric Administration
NRA	National Recovery Administration
OMB	Office of Management and Budget
OPEC	Organization of Petroleum Exporting Countries
OWM	Office of War Mobilization
OWMR	Office of War Mobilization and Reconversion

PIES   Project Independence Evaluation System
PUC   Public Utility Commission
SALT   Strategic Arms Limitation Treaty
SST   Supersonic Transport
START   Strategic Arms Reduction Treaty
TTAPS   Report by Turco, Toon, Ackerman, Pollack, and Sagan in 1983
TVA   Tennessee Valley Authority
WAES   Workshop on Alternative Energy Strategies at MIT

— 1 —

Introduction

Global warming threatens to destroy the world climate, turn the Midwest into a desert, flood Florida, Long Island, and much of the East Coast, and spawn hurricanes, fierce storms, and forest fires. Hurricane Katrina's destruction of New Orleans is just the beginning. Overseas, the ocean will flood Holland and Bangladesh, and fertile farmland in Europe and Asia will dry up. This is one apocalyptic future many environmentalists predict. Another apocalyptic danger is the population explosion, with the number of people growing from 6 billion to 8 or 10 or 20 billion, far outstripping arable farmland, oil supplies, and forest resources. A third is the shortage of energy. Even President George W. Bush has spoken of an energy crisis. The gravest danger, however, comes from nuclear warfare. The Americans and Russians have about 10,000 atomic missiles. Exploding even a few hundred could trigger nuclear winter, where dust and smoke would prevent sunlight from penetrating the atmosphere, leading to starvation and the death of the human race. A few atomic bombs from a rogue state like Iran or North Korea could trigger a final Armageddon.

Environmentalists have spoken in apocalyptic terms for many years. Rachel Carson warned, "Pesticides will mean that no birds will sing in the Spring."[1] Bill McKibbin predicted "the end of nature."[2] Others proclaim, "We are destroying the earth," "the population explosion will end civilization," and "global warming will make life impossible on the planet." In a cover article, the *Atlantic Monthly* announced "The Great Climate Flip-flop," predicting, "We could go back to ice-age temperatures within a decade...."[3] Al Gore wrote that the dangers were so great that the future of the earth "hangs in the balance," and more recently produced and starred in a popular documentary film about global warming. Such rhetoric is a standard feature of the movement. Perhaps the most influential was the Club of Rome report issued in 1972, titled *The Limits to Growth,* which forecast that a combination of pollution, overpopulation, and energy shortages would cause the entire social and economic system of the world to "crash" about the year 2020.[4] Ominously, the very next year the energy crisis began, touched off by the Arab oil embargo.

Couching forecasts in apocalyptic terms derives, of course, from the Bible, particularly the Apocalypse of St. John, and earlier the Book of Daniel. Modern prophets of environmental doom are both shaped by the biblical tradition and fascinated by its persuasiveness. In terms of its narrative power, few stories can surpass the Revelation to John, to use its other name. Christ returns to destroy the old, sinful world, awakens the dead, establishes his heavenly kingdom, and in the climax, drowns forever the devil and his evil followers in a lake of fire. This concept of total destruction of the world appeals powerfully to environmentalists, who, like the biblical prophets, foresee the end of the earth. Though usually not accepting the religious concept itself, they like the doomsday rhetoric and images.

The Bible defines the Apocalypse as truly the end of the world. In this book the definition is slightly less devastating: an end to civilization, or a state of continuous warfare, or an end to the material prosperity of the industrial West. Common sense dictates special preparation to avert these apocalyptic threats. Regrettably, over the last 50 years and longer, Americans have typically ignored the dangers or have engaged in furious planning for a few months or few years but then fallen away from their good intentions.

On a lower level of threat, the United States and the world face many other dangers like the AIDS epidemic, extreme Muslim terrorism, racial conflict, and hazardous waste that—serious though they may be—do not threaten global destruction. They have less potential for total devastation and, presumably, can be reversed. Lower down this hierarchy of doom are dangers like bad schools, urban sprawl, or censorship that obviously can be corrected if people have the political will.

Threats of energy shortage, overpopulation, nuclear war, and global warming are all international. The Club of Rome forecast the crash of the world economy. The oil crisis harmed every industrial and semi-industrial country. Because oil is easily transported and is essentially the same everywhere, all importers suffered and all exporters gained. For the same reason—the global scope of the economy—overpopulation in China, India, or Nigeria will put pressure on all countries. The destruction caused by nuclear war is obviously international. Global warming threatens all countries with drought and rising sea levels.

Within the United States, policymaking for apocalyptic threats departs from the typical pattern. For ordinary domestic issues Congress plays a key role, but in international relations the president and the State Department dominate, and Congress and the courts defer. Apocalyptic threats also give a more prominent role to scientists and other experts. The Club of Rome report was entirely the creation of experts and could not be undertaken until the development of mainframe computers. The threats of overpopulation and high prices for oil and gasoline, on the other hand, can be appreciated by ordinary citizens, but understanding their timing and form depend on experts. Nuclear destruction is easy to understand, but knowledge of the technology and risk is limited to experts. Moreover, the experts in the Pentagon conducted much of their planning in secret and even misled the public. Global warming cannot be observed or measured by nonscientists and is a phenomenon many years in the future. Depletion of energy, minerals, and land, overpopulation, nuclear war, and global warming are all apocalyptic threats apparent at the present. They are apparent, however, only because modern people believe that the future will not necessarily be the same as the past, and because scientists and statisticians have gathered and analyzed information about the problems.

PLANNING

The roots of societal planning date from the eighteenth-century Enlightenment. American founders like Franklin, Jefferson, and Madison were self-conscious in planning a national government and the country's orderly expansion into the Northwest Territory and beyond. The French revolutionaries planned a radically new society. Slightly later, Saint-Simon and Comte planned less radical societies. Starting in 1804 the general staff of the Prussian Army developed the techniques that continue to dominate war plans.

Within the United States, planning did not flourish until over a century later. In 1929 President Herbert Hoover appointed the Research Committee on Social Trends. Then in 1933, with Franklin Roosevelt's New Deal, the floodgates opened. The best example still in existence is the Tennessee Valley Authority, which included programs for agriculture, nutrition, education, and recreation along with hydroelectricity, navigation, and flood control. The most sweeping central planning, however brief its sway, was the National Industrial Recovery Act, passed in 1933 and struck down by the Supreme Court two years later. The National Recovery Administration set the country on a course of detailed collective planning with similarities to the Soviet five-year plans and plans by the Fascists in Italy and Nazis in Germany. The NRA imposed control on virtually every sector of American economy, requiring each industry to establish management-labor teams that set production quotas and prices. But when the Supreme Court declared the law unconstitutional, Roosevelt abruptly reversed course and abandoned collectivism. With the outbreak of World War II in Europe, however, mobilization began, putting the economy and society under pervasive control from Washington.

After victory in Europe and the Pacific, President Truman and Congress rapidly terminated the war mobilization. But overseas, planning emerged again once the United States began the Marshall Plan of aid to Europe. The program required each country to create a central economic recovery plan. Later in the 1950s and 1960s, U.S. aid sent to underdeveloped countries also required them to engage in comprehensive planning. In this period the Agency for International Development began population control programs abroad. Lyndon Johnson's Great Society called for planning in its model cities programs and initiated federal regional councils to plan and coordinate.

Richard Nixon, often an enigma, embraced central planning, even though Republicans generally rejected it. Nixon's most dramatic intervention was his attempt to control all wages and prices in 1971. His legislative authority was the old World War II law on war mobilization, dusted off and revised early in his term. Within months of the wage and price freeze, it became obvious that the Treasury Department's predictions were completely erroneous, so Nixon, always a realist, abandoned the scheme. Two years later, however, when the Arab oil embargo and OPEC price rise brought shortages of gasoline and heating oil, Nixon quickly appointed an energy czar and established the emergency Federal Energy Office in the White House, which began planning and allocating petroleum products. Soon Congress passed the Petroleum Allocation Act to give it legal authority. Like the freeze, planning for petroleum soon showed its inadequacies. For the longer term Nixon established Project Energy Independence. In writing the Clean Air Act and Clean Water Act in 1970 and 1972, Congress embraced the New Deal heritage of central planning. The laws required state plans for air pollution and river basin and metropolitan-wide planning for water pollution. The Coastal Zone Management Act and other laws passed during the 1970s required comprehensive planning as well.

While the Carter administration marked a high point for central government planning not seen since World War II, Ronald Reagan was its sworn enemy and advocated free market solutions. Although the Reagan administration partially rolled back planning and central control, it did not eradicate it, and its supporters in Congress and the bureaucracy bided their time. The Bush (Senior) administration was more compatible with these approaches, and President Clinton was positively eager. By his own characterization, he was a "policy wonk," enthusiastically debating detailed issues and proposals.

Because an apocalyptic threat is worldwide, domestic planning is inadequate without international cooperation. In negotiating treaties on population, disarmament, and global warming, the United States deals with countries with a stronger tradition of planning. The Europeans and Japan have long embraced planning and find it frustrating that the Americans are not so enthusiastic. To them it seems ironic that the country that demanded their formal planning in the Marshall Plan and Japanese recovery now is so chary.

THE APOCALYPTIC TRADITION IN AMERICA

American political rhetoric has a powerful legacy from biblical prophecy, starting with the Pilgrims. Until well into the twentieth century, nearly everyone read the Bible regularly and heard it preached on Sunday. Even the humblest log cabin on the frontier had its copy, perhaps the only book the family owned. The prophetic tradition energized the Revolution and the antislavery movement. Abraham Lincoln was a master at biblical references. Later in the nineteenth century, orators like William Jennings Bryan roused their audiences with biblical allusions. Today public opinion polls show Americans to be the most religious people in the world. Seventy percent say that they are believing Christians, and 40 percent believe in an apocalypse. The *Left Behind* series has sold millions of books and inspired three feature films.

The Christian Bible ends with St. John's dramatic vision of the destruction of the world. When writing in A.D. 90–95, John modeled his Revelation on the Book of Daniel. The *Cambridge Bible Commentary* defines an apocalypse as a particular form of prophesy that is written rather than oral and has a pseudonymous author, supposedly an ancient man of God. It claims to disclose secret information about the future known to God but hidden from man and uses symbols and fantastic figures. It claims that the end will come soon. The predictions often come in a vision or dream that needs to be interpreted. An apocalypse predicts catastrophe; Daniel forecast the devastation of the Jewish people and their Temple, and John forecast the end of the earth. If it is not the absolute end of time, an apocalypse predicts the end of an era.

The Bible is filled with mini-apocalypses; indeed some are environmental. In Genesis 3, God expels Adam and Eve from the Garden of Eden. In Genesis 6, God tells Noah to build an ark to save his family and all the animals. Moses, Jeremiah, and Ezekiel are prophets of apocalyptic dramas. The apocalyptic masterpiece of the Old Testament is the Book of Daniel, which the author claims is set in about 600 B.C. and consists of predictions of the next four and a half centuries, which was much easier to do in the year 164 B.C., its actual date of composition, when its real purpose was to exhort and support the Jews in their struggles against the persecutions of Antiochus Epiphanes. Mark and Matthew tell how Jesus predicts the end of the world, preceded by war, famine, and earthquakes. Specifically referring to Daniel, he goes on to say that at its climax "the sun will be darkened, and the moon will not give its light, and the stars will fall from heaven...." Standing across the valley from Jerusalem, Jesus looks at the Temple to predict its destruction, saying, "there will not be left here one stone upon another, that will not be thrown down."[5] Mark's account is known as the Little Apocalypse. Like the author of Daniel, the authors of the Gospels had the advantage of hindsight, since the Book of Mark was written about A.D. 70 and Matthew was written in the 80s, after the Romans destroyed the Temple in A.D. 70 to punish the Jews for their rebellion. John reworks the Book of Daniel and the eschatological Gospel passages into the apocalyptic masterpiece that concludes the New Testament. The author reveals secret knowledge obtained in a vision, elucidates a battle between good and evil, foretells divine judgment, and predicts the end of the world.

Augustine, writing after Christianity became the official religion of the Roman Empire, treated Daniel and Revelation as allegories. Church doctrine during the Middle Ages followed Augustine in considering the Apocalypse as an allegory. Early Protestants, on the other hand, elevated the Apocalypse because it foretold a revolution. Although the growth of science during the Enlightenment undercut belief in the Apocalypse among intellectuals, it still appealed to conservative theologians. Joseph Mead, a seventeenth-century Anglican, took it literally and calculated the dates it seemed to predict. Other theologians joined this approach and combined it with elements of the Enlightenment to conclude that the millennium would not come as a single divine intervention, but with the gradual unfolding of history.

Humanity would progress into the heavenly kingdom in stages, giving it the name of progressive millennialism.

This appealed strongly to Americans. In New England Jonathan Edwards introduced the idea, offering the discovery and colonization of America as proof. The patriots adopted the concept, identifying the 13 colonies with ancient Israel and recycling Reformation rhetoric against the pope in the argument against King George III in the Declaration of Independence. Nineteenth-century preachers and politicians asserted that it was the new nation's manifest destiny to settle the West and eventually take Hawaii and the Philippines. Abolitionist societies dedicated to freeing slaves considered their work a continuation of God's plan. "The Battle Hymn of the Republic" sings of Christian progress. By 1894 the Reverend Walter Rauschenbusch gave the name social gospel to a movement to raise up the poor and uneducated in the slums of New York. He saw it as the last stage before the second coming of Christ.

Theodore Roosevelt concluded his famous 1912 speech to the organizing convention of the Bull Moose Party: "We stand at Armageddon, and we battle for the Lord." Woodrow Wilson spoke frequently of carrying out God's will and entered World War I in order to make the world safe for democracy. The war itself was frequently compared to John's Battle of Armageddon. General Dwight Eisenhower titled his book on World War II *Crusade in Europe*. Hitler's murder of six million Jews is now called the Holocaust, a word referring to a burnt offering in a temple. Leaders of the United States and western Europe defined the Cold War as a battle between freedom and slavery, and between good and evil. President Reagan called the Soviet Union the Evil Empire, and George W. Bush echoed this when he spoke of an Axis of Evil. From 1950 to 1990 the possibility of thermonuclear destruction, truly an apocalyptic event, dominated international politics.

THEMES

This book is not about religion, however, but about politics, government, and science. It asks why the apocalyptic rhetoric is so strong. Many who use it are scientists, and the choice may seem strange since many are not personally religious and since scientists are famous for claiming to adhere strictly to the facts. Certainly many of the gloom-and-doom predictions are attempts to persuade. The prophecies are intended to alarm the public and politicians in order to change policies. But this is not entirely manipulation. Most of the alarmists sincerely believe that overpopulation will lead to war and famine, or that nuclear war is likely, or that global warming will destroy the earth. A less religious form of apocalyptic imagery is to allude to war. Military planning for nuclear war is the ultimate expression. Many advocates of population control warn that overpopulation will lead to a war over resources. Both Nixon and Carter compared the energy crisis to war. Prophecy is integral to an apocalypse. The modern prophet, however, does not wear a hair shirt and cry in the wilderness, but comes armed with computer printouts and scenarios to predict the future.

The theme of science and statistics looks at how prophecy is now couched in terms of numbers. No modern Jeremiah would merely predict doom and exhort people to change their ways. Now he or she parades a stack of statistics, filtered through complex computer models. Perhaps the most dramatic and influential forecast was by the Club of Rome, which was the first to use the mainframe computers and feedback analysis. Besides the assumption that numbers are the way to persuade, modern visionaries look to paradigms of military planning, such as World War I. Their civilian equivalents may be found in New Deal programs to fight the Great Depression and in their European counterparts, especially in Italy. These plans were self-conscious and sought to be comprehensive. They were not necessarily internally consistent.

The four apocalyptic problems of energy, overpopulation, nuclear war, and global warming are characterized by having a life cycle, or often several cycles. Global warming, for instance, was first recognized in 1896, then dropped from sight to reemerge in 1985. Energy shortages were predicted in 1909, 1952, and 1956, then ignored until 1973. While warnings of many apocalyptic threats come from individuals, in the complex world of science, government commissions often give the alarm. Once a problem is recognized, the national government almost always takes the lead. This means the president and Congress. Yet sometimes top political leaders feel an issue is too controversial and so duck it. In 1959 the Draper Committee told President Eisenhower explicitly and clearly about the danger of overpopulation, but Ike did not want to hear about it. He ignored the issue for the rest of his term in the White House.

After a threat is recognized, the next stage is to prepare a plan to deal with it. For example, President Carter offered his national energy plan, and Eisenhower proposed massive retaliation to defend against Soviet missile attack. Such plans are self-conscious, but not always comprehensive or consistent. Plans may be international, such as the Kyoto Protocol or the Nuclear Non-Proliferation Treaty. In the four threats discussed here, the international negotiations were largely shaped by the United States, even in the case of Kyoto, which was later rejected. From 1933 on planning has focused on the president directly. Furthermore international negotiations are under the thumb of the president. Earlier cases focused on Congress, such as the 1924 law on immigration that had the goal of maintaining the existing ethnic mix. After a plan is designed, supposedly the next stage is to implement it. This is only in theory, however, for frequently plans are ignored even when they appear good in terms of goals and methods. For example President Ford rejected Project Energy Independence within weeks of its official announcement. A final theme of the book is the extent to which a plan is determined by a similar prior situation. This was the case for the Kyoto Protocol, which copied the Montreal Protocol on controlling ozone.

NOTES

1. Rachel Carson, *Silent Spring* (Boston: Houghton Mifflin, 1962).
2. Bill McKibbin, The End of Nature (New York: Random House, 1990).
3. William H. Calvin, "The Great Climate Flip-flop," *Atlantic Monthly,* January 1998.
4. Donella H. Meadows, Dennis L. Meadows, Jorgen Randers, and William W. Behrens, *The Limits to Growth* (New York: Universe Books, 1972).
5. Mark 13 and Matthew 24:29 and 24:2.

— 2 —
The Golden Age of Statistics: Planning as Prediction

Predicting an environmental catastrophe is the first stage in planning to avoid it. One of the most successful and controversial predictions was the Club of Rome report in 1972. Its importance lies in its being the first massive computer simulation of a coming apocalypse. Its input was statistics on arable land, energy, minerals, population, pollution, and economic growth encompassing the entire world. Since then other comprehensive analyses like *Global 2000* and the Brundtland Commission have repeated the Club of Rome effort. The 1972 report was a warning, a forecast, and a prediction—one that took the world by storm, selling millions of copies in 29 languages. While the report was a tour de force and raised planning to a new high, systematic planning has been around for at least two hundred years and has antecedents dating back to antiquity.

Archeologists digging in Sumer have found evidence of oneiromancy (predictions based on dreams) as far back as 2500 B.C.[1] The Book of Genesis describes how Joseph interpreted the pharaoh's dreams to forecast plagues and floods. The Book of Daniel, the first of the two great apocalyptic books, describes how Daniel interpreted the dreams of King Nebuchadnezzar to predict the fall of Babylon. The ancient Greeks and Romans used pyromancy, scapulimancy, and tyromancy to forecast the future from fire, shoulder bones, and cheese. Even today, newspapers print astrological guides. These venerable methods are now recognized as phony, at least by educated people. Daniel's incredible accuracy is bogus, having been written about 165 B.C., not three hundred years earlier as it claims. The Revelation of John shares fewer of these characteristics, because, according to modern scholars, much of it is an allegory attacking the Roman government by equating it to Babylon and evoking the Book of Daniel. Revelation extends farther than Daniel in systematically describing a future utopia of the New Jerusalem.

Systematic forecasting did not emerge until the modern age. Of course, farmers back to the Stone Age could forecast the annual cycle of seasons, and generals always have tried to predict realistic outcomes of their military campaigns. The ancient Sumerians, Hebrews, Greeks, and Romans could foresee the danger that enemy armies would attack their cities, so they built walls. Largely, however, as long as people believed God or the gods ruled the world, predictions were divination, literally discovering the divine will. Only when the age of faith gave way to the Enlightenment could forecasting occur, because only then did people believe in a future.

UTOPIAS AND VISIONS OF THE FUTURE

The Renaissance produced a spate of visions of the future, the best known of which was *Utopia,* written by Thomas More in 1516. Many of the visionaries were French. In 1771 Sabastien Mercier published *The Year 2440,* describing "the halcyon days of France" nearly seven centuries in the future. Citizens live in an ideal social system, with prosperity and education available for all. Mercier believed that sooner or later this was inevitable. Mercier's younger contemporary, the Marquis de Condorcet, held similar ideas and believed enthusiastically in progress. The discovery of America, with its primitive society, pointed toward the idea of progress over time. He maintained that human nature was perfectible (intellectually, morally, and physically). Henri de Saint-Simon advocated that scientists should assume the role of leaders, becoming a new priesthood. His disciple, August Comte, envisioned an ideal state led by sociologists.

American founders like Franklin and Jefferson (who both lived in France for many years) also had a vision of the future. The founders believed that education, justice, and the end of hereditary privilege would produce the new type of citizen. Their knowledge of social trends was rudimentary. For instance, John Adams wrote how the growing population would provide soldiers for the Continental Army, noting that each year 50,000 young men would reach the age of 18, more than enough to replace casualties and deserters. The founders were aware that the growing population needed the outlet of farm land in the west for settlement. The Northwest Ordinance enacted under the Articles of Confederation incorporated social as well as topographic planning. The land was to be sold to promote individual ownership of farms, and each township was to reserve one section to support a school. The Constitution of 1787 provided for a census, which generated the first national statistics.

While Mercier, Condorcet, Saint-Simon, and Comte represent the utopian strain of enlightened French thought, the physiocrats sought a better future, but without detailed planning. The founder of this first school of economics, François Quesnay, believed that all value could be traced to land. The route to prosperity was absolute free trade, and the proper governmental role was to do nothing, or "laissez faire." The Scottish political economist Adam Smith learned of these doctrines while traveling in France and incorporated them in *The Wealth of Nations*. In contrast to the physiocrats, he considered the key to be labor rather than land, but like them he advocated laissez faire to transform the multiple selfish goals of individuals into public welfare. Unlike the physiocrats, Smith recognized the practical problems such as monopoly, which he opposed. Laissez faire theory had no place for planning or prediction.

The British reformer and early social scientist John Stuart Mill admired Condorcet, St. Simon, and Comte and was personally acquainted with Jeremy Bentham, the founder of utilitarianism. Mill's *Principles of Political Economy* was the most popular text on that topic for 30 years. One trend he frequently asserted was that underdeveloped regions like India would eventually follow the route of the advanced ones. Moreover, the advanced countries would continue to make economic and social progress. This would derive from man's power over nature based on scientific knowledge. Mill believed that economic growth would not be overtaken by population growth, so per capita incomes would rise.

Timeline for Statistics

1087	William the Conqueror orders the Doomsday Book.
1666	Complete enumeration of New France (Canada).
1790	United States takes first national census anywhere.
1798	Malthus publishes *An Essay on the Principle of Population*.
1804	Prussia establishes its Great General Staff and begins war games.
1834	Royal Statistical Society founded in London.
1839	American Statistical Association founded in Boston.

1856	International Congress of Beneficence, a predecessor of the International Labor Organization, meets in Brussels.
1869	Massachusetts establishes the Bureau of Statistics of Labor.
1885	Congress establishes a national Bureau of Labor Statistics.
1890	U.S. Census uses punch cards for sorting with an electrical tabulator.
1902	U.S. Census Bureau becomes permanent.
1908	Pres. Theodore Roosevelt convenes the White House Conference on Natural Resources.
1917	War Industries Board fixes prices and sets quantities for industry.
1919	International Labor Organization founded.
1921	Soviets establish the State Planning Commission.
1922	Italian Fascist, Benito Mussolini, marches on Rome, becomes premier, and introduces corporativism.
1928	First Soviet five-year plan begins.
1927	Aurelio Peccei joins the Fiat Corporation in Turin, Italy.
1929	Pres. Hoover establishes the Committee on Social Trends chaired by William Ogburn.
1933	Congress passes National Industrial Recovery Act. New Deal agencies plan and forecast.
1942	Brig. Gen. Eisenhower begins planning to invade Europe and defeat Nazis.
1945	European recovery planning begins. Peccei helps rebuild Fiat after the war, then goes to Argentina.
1946	Full Employment Act regularizes Keynesian economics.
1947	Marshall Plan requires European countries plan their economies.
1952	Materials Policy Commission, chaired by William Paley, recommends energy planning.
1954	Peccei begins to cry the alarm over the Predicament of Mankind.
1966	Peccei travels to United States to discuss the Predicament.
1968	Club of Rome founded.
1972	Club of Rome reports the world economy and environment likely to crash in 50 years.
1973	Oil embargo by Arab countries.
1975	Project Independence predicts United States will be able to export oil by 1985.
1979	Margaret Thatcher becomes prime minister of UK with a free market policy.
1980	*Global 2000* forecasts environmental problems in following 20 years.
1981	Reagan administration rejects government planning in favor of a free market.
1983	Brundtland Commission forecasts overpopulation and shortages of resources.
1990	Collapse of USSR and socialism discredits planning.
2001	Vice President Cheney issues energy plan.

Karl Marx shared a belief in economic and social progress with his fellow reformers of the nineteenth century. He observed the process of industrialization and believed it would continue. In particular he foresaw continued technological improvements. Marx observed that the discovery of America and the rounding of the Cape of Good Hope had opened up fresh ground for exploitation. Modern industry established a world market. "The need of a constantly expanding market for its products chases the bourgeoisie over the whole surface of the globe. It must nestle everywhere, settle everywhere, establish connexions everywhere.... It has agglomerated production, and has concentrated property in a few hands. The necessary consequence of this was political centralisation."[2] Marx anticipated that big firms would squeeze out small ones. Capitalists would increasingly oppress the workers, wages would decline, and misery would increase. Depressions would afflict the economy. Eventually the workers would organize a revolution to overthrow the system. Marx's predictions about the future after the revolution were vague. While he believed fervently that capitalism would be overthrown in favor of socialism and later communism, he was unclear about its details. Marx disdained the utopian writings of the British and French reformers, including the socialism of St.-Simon. He condescendingly wrote that "it became the vocation of the aristocracies of France and England to write pamphlets against modern bourgeois society."[3]

In contrast to the visionary writers, city planning was a practical utopianism that emerged at the turn of the twentieth century. In the United States, the city beautiful movement built monumental buildings, tree-lined boulevards, and grassy parks. Marble courthouses and city halls still dominate many of their downtowns. Contemporaneously the conservation movement recognized the importance of planning for forests and agricultural land. Theodore Roosevelt and his chief of the Forest Service, Gifford Pinchot, were leaders. In 1910 Pinchot accepted an invitation from a freshman New York state senator, Franklin D. Roosevelt, to address the Committee on Forests, Fish and Game, which he chaired. Pinchot dramatically projected slides of a village in China, where antique paintings showed fertile farms, in contrast to recent photographs of the same village showing eroded and depleted fields that condemned the inhabitants to poverty and starvation.[4]

STATISTICS

The free-flying imaginations of visionaries need substance, such as practical numbers and facts about the state. Ancient civilizations like the

Egyptians and Greeks conducted censuses. Luke dated the nativity of Jesus in terms of a census that, typical of Roman government, was to determine taxes. In 1087 William the Conqueror gathered accurate information on England by sampling one man in each hundred, including his family and property. The compilation gained the nickname of the Doomsday Book because of its unappealable authority in cases of disputes was as certain as God's judgment at the end of the world. The connection to taxation, as well as to military conscription, made taking a census unpopular with both ordinary people and local officials. In 1666 the French colony of Canada conducted a complete enumeration, and in 1749 Sweden took a census based on parish registers. In France, royal attempts to conduct a census were blocked because of local fears of increased taxes, and in Britain Parliament voted in 1753 against a census because it was a threat to local government. Some ordinary people believed it violated the will of God, being the sin of David, a reference to the books of Samuel and Chronicles, where the ancient king is punished for conducting a census.[5]

Explorers of North America reported data on the land, natives, minerals, and so forth. A census of Virginia in 1624 found 1,202 English inhabitants, and one in 1635 found 4,914. The unexpectedly low numbers suggested to the sponsors that the colonization was not proceeding well. The New England colonies carefully recorded their births and deaths, mindful of the Old Testament injunctions to account to God. All colonies kept parish registers. By the 1660s, the British had recognized they had made mistakes in planting colonies and the need for more careful planning, so the sponsors tried to work out the legal and administrative structure in advance of establishing more colonies. William Penn found an investor and advisor for his colony in William Petty, who urged him to collect exact information on arrivals, births, deaths, and trade.[6]

The science of statistics dates from 1632 when John Graunt began to analyze reports of deaths in London. Finding that living in London was unhealthful, he concluded that one cause was air pollution due to burning coal.[7] Using the term "political arithmetic," his colleague (and Penn's advisor), Petty creatively used ratios and weighted means to estimate population size, agricultural production, trade, and tax revenues. Graunt and Petty's younger contemporary (and the discoverer of the comet), Sir Edmund Halley, calculated the population of England at five and a half million, the first reliable estimate ever. Contemporaneously Germans were developing "statistik," meaning facts about states arranged in tables. Although that terminology became preferred, originally their facts were primarily not numerical.[8] Indeed, they were more like political science, for example an entry on the table for type of government might be "kingdom," "duchy," or "republic."

In 1798 Thomas Malthus used rough statistics when he published *An Essay on the Principle of Population*. Combining data on population increases with data on the availability of farm land, he predicted that people would starve to death because their numbers would increase faster than the supply of food. Data from North America showed a high birth rate, which could not be sustained in the British Isles or Europe where new land was not available. Malthus argued that the two trends would collide and drive down wages to the level of poverty and famine. This appeared to match the contemporary situation in England, Ireland, and France, but contradicted the more optimistic conclusions of Condorcet. Regrettably the quality of the data available to Malthus was poor.

Although retarded by inadequate mathematical techniques, many governments began to take censuses. The United States led the way in 1790 with the first modern national census anywhere. The immediate requirement was to determine the number of representatives to which each state was entitled in the House of Representatives. More broadly the Constitution recognized that as the country grew, the ratios would need to be adjusted every 10 years. Sweden began a periodic census 10 years later. Britain took its first decennial national census in 1801, and France took one in 1821. Physicians used statistics to describe the conditions in orphanages, prisons, and mental hospitals. One study analyzed the efficacy of bleeding for treatment of pneumonia, demonstrating that it did not work. Other reformers examined the impact of education, finding for example that 67 percent of criminals were illiterate.[9] The cholera epidemic that reached France in 1832 challenged physicians to use statistical analysis, leading to the establishment of the Statistique Generale de la France, a government agency.[10]

The mathematical basis of nineteenth-century statistics was weak, so even after a census was taken, it was unclear how to use the data. The Belgian astronomer and mathematician Adolphe Quetelet attempted to apply methods of the natural sciences to social issues, winning fame for two ideas: the concept of the "average man" and understanding distributions. He formed his hypothetical "average man" based on the records of army conscripts, which tabulated the height, weight, age, and health of thousands of new soldiers. The complement of the average man was the normal distribution. Quetelet recognized how this bell-shaped curve would solve the problem of determining when a population was homogeneous, a difficulty when using census data. Developing laws of probability, Quetelet correlated variables such as between the birth rate and rural or urban locations.

The year 1834 marked the formation of the London Statistical Society, later called the Royal Statistical Society. Five years later, statisticians in Boston organized the American Statistical Association. The first International Statistical Congress met in Brussels in 1853, under the leadership of Quetelet.[11] Nineteenth-century reformers used statistics to advocate improvements in labor conditions. A series of international conferences eventually let to establishing the International Labour Organization (ILO). Woodrow Wilson was an enthusiastic supporter, serving as vice president of the American Association for Labor Legislation, a position he continued to hold even while in the White House. In 1919 the ILO scheduled its first meeting in Washington, DC, but problems occurred. Wilson suffered a stroke and was incapacitated. For a while the conference could not even find a place to meet. Assistant Secretary of the Navy Franklin D. Roosevelt came to the rescue with help from the Navy.[12]

STATISTICAL AGENCIES

The U.S. census expanded from its minimal function of counting the number of citizens in order to apportion the House of Representatives. The 1810 census asked about manufacturing, that of 1820 asked about fisheries, and that of 1850 asked sociological questions about church membership, pauperism, crime, and taxation.[13] While each decennial census gathered more and more information, the negative effect was that the amount of data overwhelmed the staff to the extent that it took nearly a decade to report the 1880 results.

The mass of data stimulated the Census Office to invent an electrical tabulator, so it hired Herman Hollerith, a young engineering instructor at the Massachusetts Institute of Technology. For the 1890 census, clerks punched holes in a card for each of 62 million individuals. When fed through a press, metal pins made an electrical contact for each hole, and the result was measured on dials. Another machine could sort the cards for further analysis. Hollerith's machine could cross-tabulate, for instance, the number of whites and blacks broken down by occupation. In 1944, the bureau recognized the potential of electronic computers, specifically ENIAC, developed by the army to calculate the trajectories of artillery shells. In the period that followed, the bureau was at the forefront of computerization.

The U.S. Bureau of Labor Statistics was another agency concerned with gathering and analyzing data. Its origin traces to several state agencies. Massachusetts had established the first Bureau of Statistics of Labor in 1869.

Its first director, Carroll Wright, claimed that he administered the bureau "as a scientific office, not as a Bureau of agitation or propaganda, but I always take the opportunity to make such recommendations and draw such conclusions from our investigations as the facts warrant."[14] In 1885 he went to Washington to head the new Bureau of Labor Statistics. His progressive policies found favor with President Theodore Roosevelt, who praised the bureau for its ability "to discover remedies for industrial evils."[15] During the New Deal, the bureau found itself in the center. Besides its usual work, the National Industrial Recovery Act and the Agricultural Adjustment Act needed vast amounts of data because they were supposed to replace market prices with administered prices. The bureau was supposed to figure out the proper levels, a task that proved impossible. When the Supreme Court found the NIRA unconstitutional after two years, that burden was lifted. World War II again brought wage and price controls, and the bureau calculated them, but it was severely criticized when its cost-of-living index became distorted during the wartime inflation.

In passing the Employment Act of 1946 Congress stated a national goal of full employment and established a Council of Economic Advisors in the White House. In 1978 Congress amended it by passing the Full Employment Act, which directed the government to undertake economic planning on a more formal basis. The council was to forecast and set goals on a five-year basis. Even President Carter considered these goals to be pious hopes, and President Reagan simply ignored them. While the power of the council has ebbed and flowed over the years, it has consistently argued against subsidies, restrictions on foreign trade, outmoded regulation, and inefficient social regulation that fails to pass the cost-benefit test.

Other agencies collected statistics on energy and minerals. The Geological Survey gathered data on land and water in the West. The Bureau of Mines collected data on mining. In 1908 Theodore Roosevelt convened a White House Conference on Natural Resources and, to follow up, appointed a National Conservation Commission to report on minerals, water, forests, and soils. The chair was Gifford Pinchot. Its inventory lacked precision, but was a first step. The following February Roosevelt convened the North American Conservation Conference with representatives from Canada, Newfoundland, and Mexico. After Roosevelt left the presidency, the commission fell on hard times and eventually had to find funds from private donors. Little happened in the following years. Later, President Herbert Hoover's Research Committee on Social Trends, chaired by William F. Ogburn, had a section on natural resources. Franklin Roosevelt had his National Planning Board gather statistics on natural resources.

Shortages during the Korean War spurred Harry S Truman to establish the President's Materials Policy Commission, chaired by William Paley, to predict world supply for the following 25 years. After 18 months of work and the help of 28 agencies, it issued a five-volume report. The Paley Commission recommended that the United States should have "a comprehensive energy policy."[16] By the time the report was issued, Truman had only six weeks left in his term, and the Korean War supply situation had improved, so little attention was paid to it. The new president, Dwight Eisenhower, was unenthusiastic, and his staff considered its backers as too tainted by the New Deal. Consequently, Paley set up a nonprofit research center, Resources for the Future (RFF), which at first he funded personally and later persuaded the Ford Foundation to fund. RFF continued to make long-range projections as connected to population growth and economic conditions.

John F. Kennedy in 1961 asked the National Academy of Sciences to investigate natural resources. The group consisted of 13 academy members, one of whom was to be liaison: Roger Revelle, the oceanographer, who was serving as the science advisor to the secretary of the interior. Controversy arose over the volume on energy. M. King Hubbert of the Shell Oil Company predicted that U.S. crude oil reserves would peak at 175 billion barrels and begin to decline in the late 1960s, producing scarcity. On the other side, the chief geologist of the Department of the Interior predicted that reserves were 590 billion barrels and would not begin to decline for at least 30 years. Accordingly the committee did not feel any sense of urgency and ignored Hubbert's predictions (which turned out to be highly accurate). The reader was left to conclude that no energy shortage would occur. Another committee volume, authored by Gilbert White, examined the effects of world population growth and how it would increase demand. Upon completion, the academy report fell into the oblivion typical of the National Conservation Commission and the Paley Commission.[17]

President Richard Nixon, at the urging of his aide, Daniel Patrick Moynihan, established a National Goals Research staff in the White House for the purpose of giving general guidance on trends through the method of publishing an annual report. It was supposed to lay out the consequences of present social trends, forecast future developments, and measure the future impact of policies. Moynihan's enemies within the White House wrested control away from him, and as a compromise, Nixon assigned the duty to another aide, Leonard Garment. One of Garment's first acts was to consult with outside experts including Herman Kahn of the Hudson Institute. Garment believed that the best function for the goals staff was to bring forward for debate some issues ahead of their time, such as growth policy, revenue sharing, and technology assessment. It deliberately ignored current issues like civil rights and the war in Vietnam because they already were receiving enough attention. The report drew some candid conclusions such as the fact that FHA and VA mortgages, interstate highways, tax regulations, and land-use programs all contributed to massive suburbanization, and that agricultural policies contributed to the depletion of rural population. It pointed to the advantages of zero population growth for the United States.

Before the Watergate scandal overwhelmed the White House, Nixon launched the analysis for Project Independence. One of the immediate complaints after the oil crisis of 1973 was that the government lacked good information. The Department of the Interior obtained its petroleum data from the American Petroleum Institute, a private organization of the oil companies, which had an obvious bias. Congress established the Federal Energy Administration, which developed PIES: the Project Independence Evaluation System. Project Independence was the public relations name Nixon gave to an aspiration of making the United States independent of foreign energy within six years. PIES was a unified linear programming model integrating data from other agencies. This evolved into IEES—the International Energy Evaluation System—to predict energy flows world-wide. PIES showed that energy independence was impossible within six years, but did offer hope that it could be attained by 1985. In fact this too was unrealistic and was merely an attempt to twist the data to satisfy the president, who by that time was long gone. His successor, Gerald Ford, had little interest in econometric prediction.

In 1977 two private organizations released comprehensive econometric studies of energy. The Workshop on Alternative Energy Strategies (WAES) was sponsored by MIT with data from the World Bank. WAES began with independent forecasts for each country for 25 years, which were then integrated. Funds came from corporations and foundations in a dozen countries. The analysis did not include Communist countries. The workshop head, Carroll L. Wilson, stressed the need to communicate with decision makers. Its participants were chosen because they held influential jobs in government and corporations. Wilson frequently briefed congressional committees, government officials, and newspaper editorial boards.[18] In another study, the Electric Power Research Institute, an entity of the big electric utilities, forecast supply and demand for the following 50 years.[19] The Brookhaven National Laboratory, a semiautonomous division of the Department of Energy, sponsored a comprehensive 25-year analysis.[20] The three studies projected various prices for oil under differing assumptions of economic growth and for 1985 and 2000, but none was as high as the actual price in 1980 after OPEC put the pressure on.

Jimmy Carter, unlike Gerald Ford, did believe in the validity of econometric models and the importance of world population, energy supplies, and agriculture. Within four months of becoming president, he directed the State Department and the Council on Environmental Quality to study "probable changes in the world's population, natural resources and environment" to "serve as the foundation of our longer-term planning."[21] The *Global 2000* report team began by surveying government data availability, which it found comprehensive. The first step was to project the population and gross national product of all countries, since the energy and natural resource demand and pollution would derive from this. The team realized that integrating various sources of information would be difficult. For example, food projections assumed that the catch from traditional fisheries would increase as demand increased, when in fact overfishing would be likely to deplete it.

The report was not sanguine, saying, "If present trends continue, the world in 2000 will be more crowded, more polluted, less stable ecologically and more vulnerable to disruption than the world we live in now."[22] The report predicted a world population of 6.4 billion in the year 2000, close to the mark of 6.1 billion. Food production would not be able to keep up, and the number of malnourished people in less developed countries could rise from 500 million to 1.3 billion. Forests would decline by 40 percent. Water supplies would not keep up with demand, and its quality would deteriorate. The *Global 2000* team determined that it lacked the data to project demand and prices for energy. The impact on the environment would be bad. Agricultural land would suffer due to erosion and desertification, and irrigated land would suffer the most. Air pollution would increase, and 14–20 percent of all species might become extinct. Although most scientists at the time did not consider global warming to be a threat, the *Global 2000* team did. It pointed out that rising levels of carbon dioxide due to burning fossil fuels and deforestation could increase the temperature of the earth by 4–6 degrees Fahrenheit. *Global 2000* included a laudatory six-page summary and critique of the Club of Rome report. It praised the computer model for its ability to demonstrate the dynamic relationships among the variables, while acknowledging the critics' complaint that the model depended greatly on initial assumptions.[23] The greatest compliment *Global 2000* paid to the Club of Rome was its imitation.

The report encountered criticism, some from experts on natural resources. They objected that *Global 2000* was prepared in a hasty manner by staff that had bureaucratic goals such as higher budgets. They observed that the first part, a 45-page executive summary that was the basis for many of the newspaper stories, distorted the 738-page technical report. Its summary was pessimistic, while the technical report was moderate. The critics protested that the econometric models were short term and hence did not account for substitution when price increased. The report disregarded historic data. More specifically, experts disputed the *Global 2000* conclusion that forests were decreasing, pointing out that instead on net they were not, and furthermore, that most logging took place in second growth areas, not virgin forests.

In a now familiar script, the report was not completed until shortly before the president left office, and Carter's successor, Ronald Reagan, had little use for it. Thus it followed the path into oblivion of the reports of the National Conservation Commission, the Paley Commission, and Kennedy's National Academy study.

In 1983 the United Nations established a World Commission on Environment and Development to reexamine the critical issues and to formulate realistic proposals for dealing with them. Its head was Gro Harlem Brundtland, the prime minister of Norway. Among world leaders, she was unique in having risen to power as the minister of the environment. Members came from 22 countries, including William Ruckelshaus, the former head of the U.S. EPA, and Maurice Strong, the Canadian UN official who organized the first Earth Summit in Stockholm and afterward the second one in Rio, advised the World Bank, and was a member of the Club of Rome. Only eight were from industrial countries; five were scientists. The commission did not conduct its own econometric studies, but relied on existing ones. Its scope was broad, covering population, food, species, energy, industry, and urbanization.

The Brundtland Commission report was balanced, but not sanguine. Its tone and conclusions sounded like the *Global 2000* report. It predicted a world population of 8.2 billion in the year 2025, but warned that total agricultural resources could not feed that many assuming a diet of 2,000 calories per day. The report noted that many species are becoming extinct due to agriculture, deforestation, and toxic chemicals. If the Amazon forests were completely cut down, the earth would lose 66 percent of its species. Energy demand presents the greatest problem. The study predicted a 40 percent increase simply on the basis of population growth. Raising the underdeveloped countries to the level of the industrial ones would be an impossibility since it would require a fivefold increase in energy.[24] Industrial growth of this magnitude would lead to more pollution of the air and water.

Like the pattern for commissions and studies within the United States, these international reports warning of environmental apocalypses were ignored.

MOBILIZING FOR WAR AND DEPRESSION

Military planning is an ideal subject for combining statistics with a future goal. Since the Napoleonic Wars, the Prussian (and later Imperial German) general staff had drawn up plans for military campaigns. This had led to victories in the war against Denmark in 1864, against Austria in 1866, and against France in 1870–71. Unfortunately, it led to disaster in 1914.

The destruction of World War I seemed to call for new methods of government and economic organization in Europe. Four paths to collective action were communism, socialism, Nazism, and corporativism. Russia had had a Communist revolution. Britain endured an economic depression and the general strike, France was crippled by the war damage, and Germany suffered from hyperinflation that destroyed the democratic Weimar Republic, eventually leading to Hitler and the Nazis. Italy, although it was officially a victor, found its government and economy falling apart. In 1922 the Fascist leader, Benito Mussolini, marched on Rome, forced the king to appoint him premier, and under the name of corporativism, moved the country toward central planning. In theory the corporate state organized workers and employers into industrial and professional corporations, such as textiles or chemicals, that would serve as political, as well as economic, organs.

Actually implementing corporativism was slow, and it was never fully in place before the outbreak of World War II swamped the effort. From the political point of view, corporativism combined a claim to national unity with an avenue for dictatorship, and from the economic point of view, it encouraged cooperation and central planning. Mussolini promoted architecture and civil engineering, building impressive marble edifices, such as the Esposizione Universale di Roma, the Foro Mussolini, and the Palazzo della Civilta. Like the Soviets and the Americans, the interwar Italians planned giant hydroelectric projects. Southern Italy was a backward region that had not enjoyed industrialization like the north; the land was eroded and needed irrigation. The government proposed a series of dams for irrigation and electricity in Cambria, Sicily, and Sardinia. Unfortunately, rich landowners fought off the program, and only smaller projects were undertaken.

In the Soviet Union, planning reached new heights. In 1921 the revolutionary government established the State Planning Commission, known as Gosplan. In 1925 it established its first five-year plan, to start three years later. At first Gosplan studied past trends and projected them onto the future, but it soon switched to a "teleological" approach, which claimed that, due to the Communist revolution, the state did not have to be confined by past trends.

In the United States, Herbert Hoover advocated the "American System," whereby government would guide industries in order to benefit from scientific rationalization and social engineering, the goal being to raise living standards, humanize industrial relationships, and integrate conflicting social elements. The key would be cooperative institutions like trade associations, professional societies, and farm and labor groups. Another name was the Cooperative Committee and Conference System.[25] While serving as secretary of commerce during the 1920s, Hoover fostered cooperation on hydroelectric power dams, foreign trade, and housing. Upon becoming president, Hoover established the Committee on Social Trends under the guidance of William F. Ogburn, a sociologist from the University of Chicago. Ogburn choose a number of statistical indicators such as migration from the farm to the city. He considered technology to be the root of economic growth. First came an invention; next the invention produced a change in industry, whence in the economy. Then the economic change affected social institutions such as business corporations and government. Finally, this produced changes in people's beliefs, attitudes, and values.[26] For example, the tractor made farming easier, but it also decreased the demand for agricultural labor. Coming at the cusp of the Great Depression, the report would have seemed to offer a tool for governmental action. Unfortunately for Ogburn, although President Hoover liked statistical data, he was not inclined to use it to design government programs.

Hoover's successor, Franklin D. Roosevelt, was the opposite, eagerly launching programs and experiments in his New Deal. The grandest of these was the National Industrial Recovery Act, which for two years brought corporativism to the United States. Americans, however, avoided the Italian terminology or the alternative, "corporatism," which was used outside of Italy. The NIRA, which Congress passed in June 1933, organized all industries into cartels representing both labor and employers, much like the Italian system. Its administrative agency, the National Recovery Administration (NRA), helped organize the industries and draw up codes that guaranteed fixed prices and wages. A top official wrote, "Industrial planning is the conscious guidance of our industrial life by collective methods." The program covered 22 million workers. Businesses that cooperated were entitled to display a Blue Eagle. Labor and management would join for parades in the streets under flags displaying the Blue Eagle. Mussolini would have felt right at home.

Two NRA divisions took the lead in planning. The statistics division attempted to gather information such as production, wages, employees, and so forth. The task proved difficult—almost impossible. The division lacked independent data, and industry refused to hand over its own data. Furthermore, the political leaders of the NRA had little appreciation of the importance of statistics and often ignored them in making decisions. The Research and Planning Division utilized the limited data to draft codes for the various industries that set out production, prices, wages, hours worked, and so on. Industry and labor representatives were then to approve or modify the codes. Then suddenly in May 1935 the Supreme Court declared that the law was unconstitutional, and the NRA disappeared. Roosevelt and his top advisors abandoned the approach and headed the country in a different direction.

The Roosevelt administration did not abandon planning and grand schemes, just the corporatist style. The Tennessee Valley Authority continued, combining social intervention with building dams. TVA manufactured fertilizer and showed farmers how to use it. It established schools for children and taught home economics to women. Elsewhere the Works Progress Administration paid to construct public buildings, roads, and parks. Funds often went to cities that had good planning departments, staffed with architects and engineers who had plans already on the drawing boards and ready to build. Much construction was in Washington, such as Federal Triangle and the Supreme Court building. Its architecture has a massive and monumental quality featuring marble and sandstone, rounded sculpture, and Roman columns, now out of style and ridiculed as "Mussolini Modern." Electricity was a common theme of the New Deal. Besides TVA, the government sponsored rural electrification nationwide: the Bonneville Power Administration in the northwest, the Fort Peck Dam on the Missouri River, and Mississippi River Dam #1 in Minnesota. All aspects of electricity demand planning and coordination.

In the United States, after Pearl Harbor was bombed, General George C. Marshall, the chief of staff of the army, sent a telegram to Texas to the newly promoted brigadier general Dwight D. Eisenhower to rush to Washington to become the head of his planning division. The next year he sent him to London to oversee planning in Europe. Ike's ability at planning, and in getting the Allies to agree to the plans, earned him command of the invasion of North Africa and later the supreme command in Europe.

Italy survived World War II amazingly well for having been a battlefield for six years, setting the stage for its "economic miracle." The wartime industrial expansion was in the range of 50 percent, and after subtracting the destruction from bombing and combat, the net increase was still 15–20 percent between 1938 and 1947.[27] The overthrow of the Fascist regime in 1943 ended much of the American and British bombing, and its new enemy of Germany had a far weaker air force. Starting in 1943 the United States supplied Italy with aid on an emergency basis, but by 1947, it was apparent that more was needed. The Marshall Plan intensified the relief by transferring American goods to European countries according to a four-year development plan approved by each government. Partly from inertia and partly from American demands for a smaller governmental role in the economy, the corporative economy withered. When the American Marshall Plan team was unenthusiastic about a government-owned steel industry, they were mollified when the private Fiat Corporation agreed to buy the production.

The postwar era was a golden age for drawing up international economic plans. In 1944 bankers and economists from the Allies met in Bretton Woods, New Hampshire, to establish postwar institutions. The key one was the International Monetary Fund (IMF), which was to administer the system of fixed exchange rates between currencies, which lasted 27 years. Another was the World Bank, which was to lend money to financially solid projects in underdeveloped countries. These were to be state-owned enterprises. When worthy projects failed to emerge, the bank encouraged governments to prepare plans to simulate them. They also became more flexible in their standards. After the fixed-rate system collapsed in 1971, the IMF turned its talents toward loans to the third world. Again their governments were required to prepare plans prior to securing a loan.

Within the United States, the success of the Marshall Plan in reconstructing Europe pointed toward expanding foreign aid to underdeveloped countries. Europe no longer needed the assistance, and the third world was threatened by international communism, as demonstrated in Korea, French Indo-China, India, Africa, and Latin America. President Kennedy's Alliance for Progress, which pledged $24 billion for Latin American countries, required the submission of plans.

THE CLUB OF ROME

The Club of Rome was the brainchild of a visionary Italian businessman, Aurelio Peccei. Born in Turin in 1908, he took a part-time job at the Fiat Corporation to pay for his education, eventually earning a doctorate in economics from the University of Turin, where he wrote a dissertation on the Soviet Union under its first five-year plan. He spoke Russian and traveled to the Black Sea region of the USSR. Upon completing his studies, he joined Fiat full time in its international division. Despite his young age, the corporation sent him to China in 1935 to supervise construction of factories to produce aircraft for Chiang Kai-shek's air force. Chiang needed the fighters to defend against the Japanese, who already had conquered much of Manchuria. The Japanese empire soon directed its aggression against China itself, including bombing attacks on the Fiat factory. Mussolini's Fascist government was playing both sides. Officially allied with Japan, Italy nevertheless wanted an export market for its aircraft industry. Fiat continued to sell aircraft and machinery to Chiang Kai-shek for an additional five months, and the Chinese paid cash to Peccei, who returned to Turin with suitcases full of money.

The Fiat Corporation was ambivalent about the Fascist government of Italy. Giovanni Agnelli Sr., the president and major stockholder, favored Mussolini, but Vittorio Valetta, the director general, secretly opposed the dictator. Peccei did not support the Fascists; he belonged to a leftist, non-Marxist political party. Valetta encouraged this allegiance and gave Peccei job duties that allowed him to travel and secretly aid the Resistance. Near the end of the war, the police arrested and tortured Peccei, condemning him to death, but in the end merely sentencing him to prison.

After the war, Peccei led in reestablishing Fiat and then went to South America to stimulate business for the corporation. He soon spotted an opportunity to begin production of tractors in Argentina. Peccei recognized that opportunities were greater in agriculture than in automobile sales. He concluded that Argentina held the greatest potential for economic growth, then expanding 10 percent annually. Colonel Juan Peron had taken power in 1943, and the country was continuing its rapid industrialization. At the time, its development was the most advanced in South America, and it enjoyed prosperity nearly equivalent to the United States and prewar Europe. Due to wartime sales, Great Britain owed it nearly $2 billion. But rather than continuing its free enterprise system, Peron shifted it toward state control and autarky. He nationalized the central bank and established government syndicates to export wheat, beef, and raw materials and to import capital goods from abroad.

Peron's policies of government ownership and management were not unusual for the time, since in Britain the Labor Party was nationalizing heavy industry, and in the United States the New Deal had built public works projects, regulated foreign exchange, and, during the war, controlled industrial production. Peron's electoral base was the working class, not the middle class or the elite, and socialism seemed to promise the best for them. The Argentine restrictions on imports meant that a Fiat factory within the country would have an advantage. The government took ownership of nearly half the subsidiary, so it wanted Fiat to succeed. Having written his dissertation on the first Russian five-year plan, and having lived in Fascist Italy, Peccei was accustomed to government planning. He wrote a five-year plan for Fiat that meshed with the Argentine government's plan. Fiat later expanded to manufacture automobiles, airplanes, and diesel engines.

Peron's central control worked magnificently for several years, but then became less successful because it discouraged investment both locally and from abroad. Wealthy Argentines began to move their money out of the country. Foreign reserves dribbled away, the electric power network decayed, and corruption increased. As the economy declined, the working class became disenchanted with the president. The middle class and the elite had always opposed him. In 1955 the army overthrew Peron and sent him into exile for 20 years. For a while, Fiat continued its privileged position with the new government. Its production of tractors, autos, diesels, and, soon, electric turbines, matched the new regime's plans, and its established position in the country discouraged other multinational corporations from competing. Yet while Fiat was doing well, the Argentine economy and social structure was falling apart. Several army coups de états failed to revive the prosperity, and the secret war began (a series of violent attacks and kidnapping by the army as well as its radical opponents). The radical People's Revolutionary Party abducted and murdered Oberdan Sallustro, Peccei's colleague who headed Fiat's operations in Argentina.

Peccei did not confine his interests to Argentina alone. He won a bid from the Venezuelan government for Fiat to build a large steel mill, established a subsidiary in Chile, and began exporting from Argentina throughout South America. Nearly all his business was in the form of governmental contracts for large projects. After the Suez Canal Crisis in 1956, Fiat headquarters in Turin became concerned about instability, as well as opportunities closer to home. Peccei arranged for the Argentine subsidiary to sell electric-generating plants to Egypt. Later he negotiated a large contract for rural development in Iran. Peccei encouraged Fiat and other Italian corporations to establish Italconsult, a nonprofit international engineering and economic consulting firm to assist underdeveloped countries. He was always the visionary rather than the specialist and had a gift for friendship. Peccei cultivated friends in business, finance, government, and even in opposition political parties. His business strategy was to employ excellent personnel, connect to agricultural projects, and integrate Fiat's plans with governmental plans, especially in their timing. His background studying the Soviet economy and living in Fascist Italy gave him a statist perspective.

In his business and travel in South America, China, Egypt, and virtually the entire globe, Peccei became concerned with "the Predicament of Mankind." Food production was not keeping up with population; tariff barriers blocked development, and most of the world lived in poverty, with poor medical care and little education. In 1956 in Buenos Aires he established the Research Bureau for International Economic Cooperation. He personally wrote a paper titled "The Underdeveloped Countries: A Great Problem of our Time." Peccei pointed with alarm to the lack of structure for industrial investment, the huge foreign debts, and the absence of planning that hindered development. One answer was more cooperation across national borders, and he applauded the nine South American countries that signed the Treaty of Montevideo to found the Latin American Free Trade Association. Another answer was to establish ADELA, the Atlantic Community Development Group for Latin America, which was to be a syndicate of major corporations in America and Europe that pledged to invest in the continent. In the United States, he won the support of Senators Jacob Javits and Hubert Humphrey. The president of the International Development Bank supported the idea.

In 1966 Peccei was invited to give a series of lectures in the United States to business groups and at Cornell, Harvard, and MIT in which he advocated the need for comprehensive statistical analysis of the problems of the underdeveloped world. While the United States had the economic expertise and was just beginning to build mainframe computers that could undertake the analysis, he argued that any such evaluation needed the participation of the underdeveloped countries and the USSR and the Communist countries of Eastern Europe. He proposed that the Ford Foundation or another giant American foundation take the lead, but none accepted his challenge. Peccei combined his proclivities toward planning with an interest in the science of the future. Years earlier, when near the end of his graduate studies, he had lived in Paris for six months, where he fell under the influence of Bertrand de Jouvenel, a pioneer of futurology. Peccei's assignments at Fiat and Italconsult often involved long-range planning. Throughout his career, he continued to monitor the Soviet five-year plans. Fiat built an automobile factory in Moscow in the 1960s and, when Fiat joined forces with the Olivetti Corporation, sold the Soviets their first computers. Based on these ventures, Peccei concluded that the Communist system was totally inadequate for development. Since the Marshall Plan had begun in 1947, Fiat had maintained close contact with the U.S. Department of State and foreign aid agency.

The next year, Peccei appeared to have gained support from the White House for a comprehensive statistical analysis. McGeorge Bundy, President

Johnson's national security advisor, pledged support and visited European capitals to promote it. Stimulated by Bundy's backing, a preliminary group convened at the University of Sussex in England, but nothing came of it immediately. Frustrated with this avenue, Peccei tried to simulate sponsorship from the International Federation of Institutes of Advanced Study. This most rarified of associations consisted of centers for advanced science and research. The Nobel Institute in Stockholm and the Rockefeller Institute in New York invited other centers such as the Pasteur Institute, the Karolinska Institute, and the Soviet Academy of Sciences. Regrettably this did not provide the right sponsor for Peccei's project, which was more practical than fitted the style of these elite centers. In 1965 Peccei published a booklet outlining his views that the world faced dangers of overpopulation, poverty, and nuclear war, and then two years later expanded it into a book, *The Chasm Ahead.* He predicted a "tidal wave of macro-problems" and lamented that no group had the vision to tackle the entire range of the issues. Consequently, Peccei would have to create a group himself.

The Agnelli Foundation, named for the founder of Fiat, sponsored a two-day meeting in Rome of 40 concerned business and government leaders, many of whom Peccei knew personally. Alexander King, the director of the Organization of European Cooperation and Development section on science, technology, and education, helped organize the session. The formal sessions agreed on the problem, but could not see how to contribute to its solution. At dinner at Peccei's house on the last evening, an informal core of six decided to pursue the problem in a practical fashion, naming themselves the Club of Rome. They began to meet regularly every six or eight weeks in Geneva, a city convenient for their travel. At Peccei's urging, the club explored a way to research the Predicament of Mankind systematically. The club soon learned of two books by Jay Forrester, *Industrial Dynamics* and *Urban Dynamics,* which employed complex computer methodologies to problems of industrial management. Forrester was a professor at the Sloan School of Management at MIT. When the 40 members of the club met in 1970, they invited Forrester to attend. At last, Peccei had found the method for his long-sought grand statistical analysis.

On the airplane back to Boston, Forrester sketched out how he would apply the Club of Rome ideas to his systems dynamics. Factors were population density, pollution, natural resources, food, available land, health, and material standards of living. Within a few weeks, six club members came to MIT for a two-week working session to launch the project. Forrester's doctoral student, Dennis Meadows, managed the research directly and played host to the Europeans. A club member persuaded the Volkswagen

Foundation to sponsor the project with a $250,000 grant. Thus Peccei finally began his long-desired evaluation in a manner much like his business projects: set out the grand vision himself, examine practical, statistical aspects, hire the best talent, focus on agriculture, foresee the solution in government planning, and emphasize timing.

The first computer runs confirmed Peccei's worst predictions: The world was moving toward catastrophe. He began to speak out to governmental leaders and distribute advance copies of the technical report. In September 1971, Meadows accepted an invitation to speak in the Netherlands, where newspapers first learned of the project. Suddenly people all over Europe were debating the controversial and pessimistic report. To maximize its impact, Peccei commissioned a simplified version in the form of a book, *The Limits to Growth,* written by Donella Meadows, the wife of Dennis.[28] He also arranged for its translation into 29 languages. Its total sales were in the millions.

The Club of Rome report combined seven major factors to predict scenarios of how they would interact over the following 50–100 years. All of them were bad. Its standard model forecast that continuing population increases at the existing rate of 2 percent per year would overwhelm the world's resources, causing the entire system to "crash" after the year 2020. Following the crash, food would become scarce, and fuels would be close to depletion. Pollution from industry would poison the globe. The report was one of the first to publicize the risk of global warming due to increased carbon dioxide in the atmosphere.[29] The demand for forests and minerals would be greater than the natural resources available.[30] A more optimistic model, which assumed natural resources reserves were doubled, would only postpone the crash about 20 years and would lead to much higher pollution.[31] Indeed the Club of Rome analysis was grounded in the positive assumptions of no major wars or massive natural disasters. Unlimited natural resources would not solve the problem because this would lead to excessive pollution. This led to the conclusion that growth could not continue, but the report pointed out that even trying to stabilize population or industrial production separately would not be enough. Stable population with growing industry would deplete resources and increase pollution, and stable industrial production with a growing population would lead to poverty. The only way out was to stabilize both population and industry.[32]

The report pointed out the need to move expeditiously. For example the Stabilized World Model I, which offered the highest level of industrial output per capita and assumes no increase in population, was obviously unrealistic. Stabilized Model II, which assumed the birth rate would decline to two children per family immediately, showed the total population continuing to grow for another generation, because children already born would reach breeding age. The result would be a more people and less industrial output per capita. If population did not stabilize until the year 2000, it would be impossible to reach equilibrium. Food and natural resources would be inadequate by the year 2100, leading to famine and poverty.[33]

The answer to the Predicament of Mankind was a global equilibrium, with industrial production as high as the average for the world today, meaning that the American and European standard of living would decline greatly as the third world rose to the average. Certainly technological advance will be necessary and welcome to control pollution and improve health. Because growth cannot be sustained, Meadows argued that do-gooders would have to give up their vague belief that poverty in underdeveloped countries would decrease, and that the gap between the rich and the poor would disappear. That could only be accomplished by direct redistribution of wealth.[34] Meadows acknowledged that the equilibrium situation would require people to give up rights like determining the number of their children or their level of consumption. On the other hand, it would promise survival at a moderate level and offer alternative opportunities for nonconsumptive activities like education, music, and leisure.[35]

The Club of Rome report met a storm of hostility. One criticism, with an element of truth, was that the club was a secret cabal of rich businessmen. To counter this, the club publicized the names of its members (which were never actually secret), and stressed that the members came from underdeveloped and Communist countries as well as capitalist ones, and that they included economists and scientists as well as businessmen. A related criticism was that the MIT computer modelers kept their data secret. In fact, the basic information came from public sources, and Forrester and Meadows promptly cooperated in making their data and calculations available.

Many criticized the report's assumptions. They claimed that agricultural productivity could easily double or triple because farming was so primitive in the underdeveloped countries. Simply introducing a steel plow instead of a wooden one would be an example. Critics disagreed with the report's pessimistic view of innovation, claiming that human invention had always risen to the challenge. They objected to the report's characterization of pollution, believing the threat was much less. Thirdly, critics objected that it was misleading for the report to lump together the entire globe. The world was not that integrated, and the proper course was to examine the world broken up by regions. The original report, of course, was dry and technical with arbitrary assumptions and timing. *The Limits to Growth,* although much improved in readability, still suffered from the technical constraints. For example, chapter five introduced the hopelessly naive condition that the average number of children in a family had to decrease to two by 1975, three years hence.[36] Although the final criticism had little scientific basis, it was the most widespread: that the report was pessimistic. This vague, but common, criticism did not argue that this or that calculation or assumption was wrong, but that "humans can do better," or words to that effect. In both the technical and popular versions, the report emphasized that its predictions were not inevitable; rather they predicted the future if present trends continued unchanged.

The availability of petroleum, coal, and mineral resources was a central feature of the report, and it also pointed out the close correlation between energy consumption and industrial output. This insight about the centrality of energy proved prescient only a year later, when the Arab oil-exporting countries in the Middle East doubled their price for crude and embargoed the Netherlands and the United States because they had supported Israel in the Yom Kippur War, setting off what soon was labeled the energy crisis. Non-Arab members of OPEC, the Organization of Petroleum Exporting Countries, joined and raised the price of oil a total of six-fold over the coming months. The Club of Rome could hardly have found more dramatic proof of their predictions. The report was also prescient about the dangers of pollution and specifically addressed the danger of global warming. *The Limits* presented a graph of Charles Keeling's observations at the Mauna Loa observatory in Hawaii showing the increase of carbon dioxide in the atmosphere.

Surprisingly few people criticized the Club of Rome for its assumptions that governmental planning and direction were the best answer, a method labeled by its critics as "dirigisme," using the French word for directing. The report frequently speaks naively of controlling the number of children or redistributing income. While the statements are always made conditionally, nevertheless the report had a tinge of Mussolini's Italy. The early 1970s were still an era of government planning in Europe and even in the United States, where Nixon had imposed a wage and price freeze in 1971, and his officials in the Department of the Interior were drawing up master plans for mining coal based on rather far-fetched national planning.[37]

A year later, in response to the complaint that a single globalized analysis ignored regional variation, the Club of Rome asked Mihajlo Mesarvic and Eduard Pestel to prepare a follow-up to the original report by dividing the world into 10 regions and added more variables. Published two years later as *Mankind at the Turning Point,*[38] this analysis tried a more sophisticated analogy of organic rather than mechanistic growth; trees grow very tall, but they do not grow up to the sky. Population growth will at some point level off. Where *The Limits* had a chapter titled "The Nature of Exponential Growth," with graph curves pointing higher and higher, *The Turning Point* had a chapter on "Organic Growth," with graphs showing the growth of trees and animals. Being written in 1973–74, the book concentrated on the oil crisis. Its regional orientation paid off in showing how high prices for petroleum could paralyze the underdeveloped countries of Africa and Asia while being merely an inconvenience for North America and Europe. On the other hand, *The Turning Point* concluded that the 10 regions are interdependent, so high oil prices, shortages of arable land, and population growth in the undeveloped countries in time will drag down the industrial countries.

Whereas *The Limits* did not offer detailed recommendations, merely predicting the bad consequences, *The Turning Point* asserted that the world could not solve the crises by traditional means (that is by the free market and the nation state), but must move toward "a complete integration of all strata of our hierarchical view."[39] After warning against the "futility of narrow nationalism" the authors called for "a practical international frame-work." Earlier they claimed that the shortage of petroleum could not be solved by the market system. Although these warnings against nationalism and calls for international frameworks extended well beyond the statistical scope of the book, they apparently did not faze the businessmen members of the Club of Rome, judging by the positive commentary appended by Aurelio Peccei and Alexander King.

Public reaction toward *The Turning Point* was more favorable than toward *The Limits*. In particular, readers did not consider it to be so pessimistic, per-haps because they had assimilated the doomsaying of the former. Another factor, however, was that this second report did not carry out the analysis for so long a number of years. The *Turning Point* only made its projections for 50 years, therefore not giving the bad outcomes time to develop com-pletely. In spite of its claims of differing from *The Limits*, *The Turning Point* arrived at quite similar conclusions. Although it broke up the world into dif-ferent regions, it concluded that they were highly interdependent. Although it claimed an organic technique, its conclusion paralleled the exponential technique of *The Limits*. In the end *The Turning Point* never became popular because it was more difficult to understand and less original.

The Club of Rome continued to study the Predicament of Mankind. After *The Turning Point,* it commissioned about 30 more reports, about one a year. Peccei published his personal reflections in *One Hundred Pages for the*

Future, pointing with alarm to the "cancerous growth of population."[40] He organized a series of high-level meetings to continue to address the Predicament of Mankind. After publishing *The Limits to Growth,* however, the club had a hard time maintaining its momentum from the peak of attention and prestige. In the field of computer modeling, others gained skills so that MIT and the club no longer dominated. Although the 1973 oil crisis had vindicated the club, as oil became more available in the years following, its glory faded. Peccei himself shifted his focus to the potential for young scientists from the first, second, and third worlds to cooperate to address the Predicament of Mankind. Other club members did not share his enthusiasm. Peccei organized the Forum Humanum with branches in Rome, Madrid, Geneva, Rio de Janeiro, and Buenos Aires, but the movement was not successful.

The club had always been a personal creation of Aurelio Peccei. He ran it out of his head from his office at Italconsult or wherever he was. The 1972 report was his personal vision. For many years before founding the club, perhaps as far back as his ventures in China and Argentina, the idea had grown in Peccei's mind of the sort of statistical analysis he sought and the topics of agriculture, natural resources, and population. With Peccei's death, the club recognized that its informal structure was no longer possible, so it became more organized. The club continues to hold an annual conference and to publish reports. In 2004 it published a *30-Year Update* by Dennis Meadows looking back at the original report.

FUTUROLOGY, CORNUCOPIANS, AND THE FREE MARKET

At the same time as Aurelio Peccei was seeking a forum for his concept of the Predicament of Mankind, others were also addressing similar issues. This parallel movement flourished, but did not develop the statistical sophistication of the Club of Rome report. During the 1960s they coalesced around the term futurology, and in 1966 they organized the World Futures Society. The terminology itself was invented in 1949 to mean the science of the future.

The French journalist, economist, and political scientist Bertrand de Jouvenel had been writing about possible futures and their consequences for many years. Prior to World War II, he had criticized the French army strategy for being unrealistic. In the late 1950s de Jouvenel founded the Futuribles Group and the journal *Futuribles,* coining the artificial word from the combination of "futures" and "possible." He disliked the term futurology because he believed prediction was more speculative than scientific and preferred the term "conjecture." De Jouvenel considered demographic and economic forecasts to be too narrow and sought to understand great social transformations, particularly in technologically backward countries. To gain a synoptic view required breaking down barriers between academic disciplines. De Jouvenel promoted a role for social scientists of moderating and explaining coming technological changes.[41] His focus on technology brought him to examine energy policy and recommend the alternative of solar power, some years before the energy crisis.[42] Aurelio Peccei had found his ideas intriguing, having first encountered them in the mid-1930s, at which time de Jouvenel had written one book on the directed economy and another on European unification.[43]

In 1965 the American Academy of Arts and Sciences asked the Hudson Institute to sketch some alternative futures. The result was titled *The Year 2000,* written by Herman Kahn and Anthony J. Wiener. Its appropriate subtitle was *A Framework for Speculation on the Next Thirty-Three Years.* This rambling and far-ranging volume characterized the world in terms of economic development, form of government, political ideology, population, military strength, and many other criteria. For those variables that could be measured statistically, the report projected future trends until the year 2000. It estimated how these outcomes would vary under differing assumptions such as a higher rate of growth, the decline of communism, and so forth. The report examined both the scenarios with and without disruptions, including wars, both guerilla and nuclear.[44]

In comparison to the Club of Rome report, the Kahn and Wiener analysis was less focused and did not create feedback loops to investigate how changes in one factor would influence other factors. The Hudson Institute study came to conclusions nearly the opposite of the Club of Rome's. It optimistically predicted nuclear power would be abundant and cheap. Capacity would increase a hundred-fold, and the cost of electricity would decline by 50 or 75 percent. Atomic bombs could be used peacefully to dredge harbors and break up rocks for mining. Oil shale in Wyoming, Colorado, and Utah were equal to two trillion barrels, five times the existing known reserves. Kahn and Weiner were on the mark in their predictions of world population being 6.4 billion. They optimistically believed that nearly all developing counties would be able to avoid the trap of economic growth lagging population growth.[45] The industrial countries of Europe and North America would enjoy great prosperity. The problems of pollution were not even mentioned. The report enjoyed a broad audience and was frequently cited.

The Hudson Institute was largely the creation of Herman Kahn. Kahn had been employed at the Rand Corporation for many years, but had grown dissatisfied with its orientation toward military analysis. After World War II, the army air force had established a division within the Douglas Aircraft Company to continue the work of its operations research staff. After the victory, these air force officers did not want to remain in uniform, even though they enjoyed their assignments. The solution was to establish a civilian consulting company that worked on contracts from the air force. In 1948 it became independent as the Rand Corporation. Kahn joined Rand and achieved prominence for his analysis of nuclear war options. In 1961 he left to found the Hudson Institute, which had the mission "to think about the future in unconventional ways."

Cornucopians is the label given to a group of optimists who hold a rosy view of the future. Most are economists rather than natural scientists. Julian Simon, who was one of the most outspoken, asserted that "almost every trend in material human welfare points in a positive direction...."[46] This included food, health, and raw materials. The price of food has decreased all over the world, and deaths from famine are down. Life expectancy in the United States rose 30 years, from 47 years to 77 years, during the twentieth century, and more in some European countries and Japan. Simon maintained that the gains were larger in the underdeveloped countries, where it rose from 35 years in 1950 to 50 years. Accidental deaths fell to one-fourth the level in 1910. Prices of copper, iron, and other metals decreased, and known world reserves increased. Petroleum was a partial exception, which Simon blamed on the OPEC cartel using its monopoly power.[47] Simon won $576 in a famous bet with Paul Ehrlich, the biologist. Simon wagered that over a 10-year period $1,000 worth of five commodities (copper, chrome, nickel, tin, and tungsten) would decrease in price relative to the consumer price index, in other words become less scarce. In fact, at the end of the decade they cost only $424, a decline of more than half.

Simon's optimism came from rejecting the view that mineral resources are finite. As an economist, he looked at prices and availability. While it was true that industry had to mine ore with a much lower quantity of copper, mining technology became more efficient. Higher prices encouraged geologists to prospect and find new deposits. Moreover the general level of personal income was higher, so people could afford to pay more. Finally, copper got substitutes such as aluminum or plastic or glass fiber. Fifty years ago it was used for kitchen ware and telephone lines, but today these demands are much lower. Simon believed that the key was the inventive power of the human mind, which is infinite. Furthermore, he believed that a larger population would expand the number of human minds on which to draw, hence the title of his major book, *The Ultimate Resource*.

Although Simon liked to gaze into the future and to publish in journals like *The Futurist,* he was an enemy to formal planning. He believed that government planning was doomed to failure for five reasons. It was biased toward the conventional views of the bureaucrats. Coordination was impossible because of administrative divisions. It tended to recommend a bigger role for government. Analysis emphasized the bad news because it gained more attention and support. And finally, it was subject to "political warping" by the president and his top appointees. Simon said that foresight advocates made a major mistake in equating business planning and governmental planning. Businesses plan for their own activities only and thereby have to control only themselves, whereas governments try to control others. Businesses bear the consequences of their planning, since they will lose money if they plan badly. When businesses plan, they take advantage of their own specialized knowledge, whereas governments try to plan for others about whom they lack specialized knowledge.[48]

The pessimistic predictions of environmentalists greatly annoyed Simon, particularly those by Paul Ehrlich, Garrett Hardin, Lester Brown, and Al Gore. He believed that the prophets of doom were motivated by the desire for attention and power. Simon often compared the modern environmentalists to the biblical prophets who foretold hell (or hell on earth) as punishment for people's sins. The modern environmentalists gained the further advantage of fund raising. He maintained that people believed the apocalyptic predictions because of the excitement it brought to their lives, poor reasoning ability, ignorance of economics, and a desire to be morally superior. Moreover, people seem to have an innate belief that the world was a better place in the past. Simon was concerned that accepting false prophets would cause long-run disaster of social disruption and resource loss. Nineteenth-century millenarians abandoned their farms and went to hill tops to await the Messiah, whereas modern environmentalists will ruin their prosperity. Americans will waste millions of dollars cleaning up chemicals in Superfund sites that do little damage if left alone, and underdeveloped countries will suffer from malaria because they do not use small amounts of DDT.[49]

Other economists shared the cornucopian optimism. Robert Solow used his Nobel Prize lecture to point out the importance of technological progress. He shared Simon's view about substitution: "It is very easy to substitute other factors for natural resources, then ...the world can, in effect, get along without natural resources, so exhaustion is just an event, not a catastrophe."[50] George Gilder, a senior fellow at the Hudson Institute, believed that capitalism had conclusively demonstrated its superiority as an economic system. He was especially taken with computers, truly a substitution away from natural resources. Internationally he preached that "the reason to defend free nations is their role as the capitalist bearers of civilization and human progress, freedom and technology, in a world that will sink into unspeakable horror without their leadership."[51] Lawrence Summers, later secretary of the treasury under President Clinton, announced to the World Bank annual meeting that "there are no…limits to the carrying capacity of the earth that are likely to bind any time in the foreseeable future. There isn't a risk of an apocalypse due to global warming or anything else. The idea that the world is headed over an abyss is profoundly wrong."[52]

The Danish statistician Bjorn Lomborg won attention in 2001 when he published *The Skeptical Environmentalist,* a 350-page book with 3,000 citations. The author began by telling how Julian Simon inspired him to write it. Lomborg laid out a series of alarmist predictions on global warming, energy, overpopulation, and so forth, then carefully refuted them by showing that the data behind them was flawed, the trends were misinterpreted, information was distorted, and countertrends were optimistic. For example, regarding global warming, he maintained that temperature records are inadequate, that warming during the twentieth century was within the range of normal variation, that the cause is likely to be solar activity or cloud cover, that computer models are inaccurate, that the correlation between carbon dioxide and warming is erratic, and that scenarios predicting further warming are not realistic. He continued to say that even if the warming is real, the consequences for agriculture in developing countries might actually be beneficial, that sea levels will rise only a few inches, that the Greenland ice cap will not melt, that health will not be harmed, and that the warming will not produce more hurricanes and tornados. Lomborg countered the prophets of doom by observing that people can adjust to a warmer climate by planting different corps and building low dikes to protect coastal cities. At present more people die from cold than from heat.[53]

Again granting for purposes of argument that the earth might be warming, Lomborg examined the costs of reducing carbon dioxide. Under the Kyoto Protocol, without trading the right to emit, the costs for the first world would be $346 billion annually. If the protocol were modified to permit trading, the costs would drop to $161 billion annually. But even an expensive program will yield only a tiny improvement because of the lag for carbon dioxide already in the atmosphere and the increase in world population. Lomborg believed it would be far cheaper to adapt to the warmer temperatures. He argued that attempts to limit emissions may be a tool and justification by environmentalists to impose their values.[54]

The cornucopians have often engaged in boisterous quarrels with environmentalists, calling each other bitter names and questioning their rationality. Indeed they seem to enjoy the fights. The critics have exposed several weaknesses. On a number of occasions Simon blundered in his calculations, for instance in his estimate that "we now have in our hands (in our libraries, really) the technology to feed, clothe, and supply energy to an ever-growing population for the next seven billion years."[55] Then he backtracked to say seven million years, a number nearly as impossible. Although he won fame for one bet, he has carelessly proposed other bets without thinking through their implications, and then reneged. Another cornucopian weakness is vagueness. Details are missing. In particular, they do not explain what the economy would be like with vastly reduced oil, metals, and timber. Gilder sings the praises of computers, but they cannot substitute for food, clothing, and shelter. Even with the present population of more than six billion, the entire world cannot consume at the American level. A billion Chinese cannot drive SUVs and live in houses of 3,000 square feet.

Faith in the free market is a key tenet of cornucopianism. Simon, Gilder, and Solow believed it would unleash people's creative abilities, thereby eliminating shortages. Yet economists recognize that the market cannot solve problems of externalities like pollution. Furthermore the underdeveloped countries tend to be those least prepared to operate under a free market system, which depends on education, rule of law, and elaborate social structures of corporations, governments, and elections. The anticornucopian environmentalists warned that, even granting the validity of the arguments, the transition may be impossible. Multiple scarcities due to soil erosion, expensive oil, and global warming may overwhelm the process. A further aspect of faith in the free market is that statistics and prediction are no longer so important, and perhaps not even necessary. Individual corporations will plan for themselves, but there is no longer a need to plan centrally.

While faith in central planning dominated the middle of the twentieth century, this changed by the 1980s. In the United States economists began to persuade government to deregulate. As early as 1975 officials in Washington began to deregulate the airline industry, resulting in cheaper fares. Jimmy Carter strongly supported the effort and moved on to deregulate other industries like trucking. As a candidate for president, Ronald Reagan advocated more of this. Besides withdrawing the regulatory hand, the Reagan administration promoted turning over functions to private business. By the end of the decade it had taken hold in many areas. David Osborne and Ted Gaebler popularized it as "reinventing government." Government was supposed to give the overall direction to society, not do all the work. In an analogy to a boat, its motto was "steer, but don't row."[56] But in fact, the Reagan administration was not very interested in steering. The market, not government, was supposed to provide the guidance. Successive presidential administrations—both Republican and Democratic—have continued more of a market orientation, albeit with numerous exceptions.

CONCLUSION

The Club of Rome report marked the high point for comprehensive statistical analysis of the world's problems. The reports that followed, like *Global 2000* and the Brundtland Commission, copied its approach, although coming to less dramatic and pessimistic conclusions. They derived their methods from the 1972 analysis. Taken together with *The Turning Point,* PIES, and WAES, the era was the golden age of statistical prediction. The Club of Rome report did much to construct thinking on the future during the 1970s. Its prophecy of doom resonated throughout the world. It predicted catastrophe only 50 years in the future, when over-population and shortages of energy, metals, and food would bring about collapse of the modern prosperous, industrial society. While the prophecies did not come from dreams like those of Daniel and John, they came from an MIT computer, which at the time seemed nearly as supernatural. The rhetoric of the report and its popularized version sounded the alarm. It warned of war and disaster. Like the biblical apocalypses, the Club of Rome believed that the collapse was not inevitable, if only people would change their ways.

The 1972 report took information that was already available and analyzed it dynamically, looking at feedback loops, a much more sophisticated method than previous studies. The data was on hand in publications of government agencies like the U.S. Census Bureau, the Statistique General de la France, and similar agencies in other countries. The United Nations was a fertile source of data. The wealth of data owed its existence to a century and more of collection by census takers and statisticians. Scientific predecessors to Forrester and Meadows included Quetelet, Wright, and others. Much ground work was laid by scientific societies like the Royal Statistical Society, governmental agencies like the U.S. Bureau of Labor Statistics, and international agencies like the International Labor Organization and the UN Population Fund. Yet these organizations lacked the comprehensive scope needed to address the Predicament of Mankind. This took the genius of a single mind, unencumbered by bureaucratic focus on existing methods. The savior was, of course, Aurelio Peccei. His single-minded drive asked the question and demanded an answer. This centrality of a single individual was exceptional.

Looking more broadly, prediction of the future could not occur in the premodern world that did not foresee a future, but only a repetition of the past. Here the utopian writers played a key role. The long line of French visionaries—Mercier, Condorcet, Saint-Simon, and Comte—was augmented by More, Mill, Marx, and others.

Because the Club of Rome report was a prediction rather than a problem per se, the American government was not forced to respond in a practical fashion. Popular reaction was quite negative, especially from industry, which largely denied any problem existed. Of course, presidents dating back to Theodore Roosevelt and Herbert Hoover had dealt with reports from important commissions that predicted shortages of resources. A common fate for them was to be ignored, often because they were not completed until nearly the end of a presidential term. Jimmy Carter took the Club of Rome forecast seriously and commissioned the *Global 2000* team. Like many previous ones, this report did not appear until too late in the president's term to be useful. Its pessimistic tone discouraged a warm reception.

Again because the Club of Rome report was a prediction rather than a problem per se, its international effect was not direct. The *Limits* gained worldwide attention and provoked discussion at the United Nations, which established the Brundtland Commission, which in turn produced a report similar to the Club of Rome and *Global 2000*. Unlike discoveries in the natural sciences, the econometric discoveries of these studies did not lead logically to technical workshops, UN- sponsored conferences, and eventually a big international conference of diplomats. The issues have come up frequently, such as at the Rio Earth Summit, but the nations of the world have not taken the next step of a major treaty.

Measured in Peccei's terms, the Predicament of Mankind has not been solved. Population continues to grow, pollution increases, oil, gas, and coal are burned rapidly, and arable crop land shrinks. Data has not been lacking. The club report analyzed huge amounts, and following reports analyzed even more data, all coming to similar conclusions. Scientific consensus has been strong. Very few experts disagree about the general results predicting population growth and resource scarcity in the future, although they do disagree about the details and how soon the problems will occur. Even the cornucopian Julian Simon agreed that population would increase, but countered that more people would bring advantages in invention and technology. Aside from Simon, most experts believe that solving the Predicament of Mankind will be difficult and expensive. The proposed solutions range from heavy-handed central control to the magic of the free market. The ideology of laissez faire economics runs counter to planning.

NOTES

1. Mark E. Cohen, Daniel C. Snell, and David B. Weisberg, eds., *The Tablet and the Scroll* (Bethesda, MD: CDL Press, 1993), p. 5.
2. Karl Marx and Frederick Engels, *The Communist Manifesto,* ed. John E. Toews (New York: Bedford St. Martin's, 1999, originally 1848).
3. Ibid., p. 15.
4. Walter L. Creese, *TVA's Public Planning* (Knoxville: University of Tennessee Press, 1990), p. 32.
5. 2 Samuel 24:1-9 and 1 Chronicles 21:1-6.
6. James H. Cassedy, *Demography in Early America* (Cambridge, MA: Harvard University Press, 1969), pp. 45, 55, 61.
7. Paul Neurath, *From Malthus to the Club of Rome and Back* (Armonk, NY: M. E. Sharpe, 1994), p. 13.
8. Paul Starr, "The Sociology of Official Statistics," in *The Politics of Numbers,* ed. William Alonso and Paul Starr (New York: Russell Sage Foundation, 1987), pp. 12–13.
9. Theodore M. Porter, *The Rise of Statistical Thinking 1820–1900* (Princeton: Princeton University Press, 1986), pp. 157, 28–30.
10. Alain Desrosieres, *The Politics of Large Numbers,* translated by Camille Naish (Cambridge, MA: Harvard University Press, 1998), p. 85.
11. S.N.D. North, "Seventy Five Years of Progress in Statistics," in *The History of Statistics,* ed. John Koren (New York: Burt Franklin, 1970, originally 1918), p. 47n.
12. Edward C. Lorenz, *Defining Global Justice* (South Bend, IN: University of Notre Dame Press, 2001), p. 73.
13. U.S. Census Bureau, "History," *www.census.gov*/main/www/aboutus.html.
14. Joseph P. Goldberg and William T. Moye, *The First Hundred Years of the Bureau of Labor Statistics* (Washington, DC: U.S. Department of Labor, 1985), Bulletin 2235, p. 3.
15. Theodore Roosevelt, "A Letter to P. H. Grace on October 19, 1908," in *The Letters of Theodore Roosevelt,* ed. Elting E. Morison (Cambridge, MA: Harvard University Press, 1952), p. 1301.

16. Gerald O. Barney, study director, *The Global 2000 Report to the President* (Washington, DC: Council on Environmental Quality and the Department of State, 1980), p. 689. Cited as *Global 2000*.

17. Ibid., p. 692.

18. Carroll L. Wilson, project director, *Energy: Global Prospects, 1985-2000: Report of the Workshop on Alternative Energy Strategies (WAES)* (New York: McGraw-Hill, 1977) .

19. Electric Power Research Institute *Fuel and Energy Price Forecasts* EPRI Report EA-411 (Menlo Park, CA: 1977).

20. David J. Behling and Edward Hudson, *Policies for Energy Conservation* (Brookhaven National Laboratory, 1978).

21. *Global 2000,* p. 6.

22. Ibid., p. 1.

23. Ibid., pp. 9, 17, 23, 26, 27, 32, 36–37, 607–613.

24. World Commission on Environment and Development, *Our Common Future* (New York: Oxford, 1987), pp. 99, 101, 169–171 (The Brundtland Commission).

25. Ellis W. Hawley, "Herbert Hoover, the Commerce Secretariat and the Vision of an 'Associative State,'" *Journal of American History* 61 (1974): 118.

26. Wendell Bell, *Foundations of Future Studies* (New Brunswick, NJ: Transaction, 1997), p. 8.

27. Vera Zamagni, *The Economic History of Italy* (Oxford: Clarendon, 1993), p. 321.

28. Donella H. Meadows, Dennis L. Meadows, Jorgen Randers, and William W. Behrens, *The Limits to Growth* (New York: Universe Books, 1972).

29. Ibid., p. 72–73.

30. Ibid., p. 124, Figure 35.

31. Ibid., p. 127, Figure 36.

32. Ibid., p. 162, Figure 45.

33. Ibid., p. 169, Figure 48.

34. Ibid., p. 178–79.

35. Ibid., p. 180.

36. Ibid., p. 166.

37. Robert H. Nelson, *The Making of Federal Coal Policy* (Durham, NC: Duke University Press, 1983).

38. Mihajlo Mesarovic and Eduard Pestel, *Mankind at the Turning Point* (New York: Dutton, 1974).

39. Ibid., p. 144.

40. Aurelio Peccei, *One Hundred Pages for the Future* (New York: Pergamon, 1981).

41. Bertrand de Jouvenel, *The Art of Conjecture* (New York: Basic Books, 1967), pp. 248–249. Published in French in 1964.

42. De Jouvenel, "Technology as a Means," in *Values and the Future,* ed. Kurt Baier and Nicholas Rescher (New York: Free Press, 1969).

43. Bertrand de Jouvenel, *L'Économie Dirigée: le Programme de la Nouvelle Génération* (Paris:Valois, 1928), and *Vers les États-Unis d'Europe* (Paris: Bibliothèque Syndicaliste, 1930).

44. Herman Kahn and Anthony J. Wiener, *The Year 2000: A Framework for Speculation on the Next Thirty-Three Years* (New York: Macmillan, 1967).

45. Ibid., pp. 71–74, 134–136.

46. Julian L. Simon, "Bet on a Better Future," *The Futurist* 31 (1997): 18.

47. Julian L. Simon, *Population Matters* (New Brunswick, NJ: Transaction, 1990), pp. 22, 25, 71.

48. Ibid., p. 415, 411.

49. Julian L. Simon, *Hoodwinking the Nation* (New Brunswick: Transaction, 1999), pp. 111–118, 89.

50. Robert Solow, "The Economics of Resources or the Resources of Economics," *American Economic Review* 64 (1974): 1–14.

51. George Gilder, "The American 80s," *Commentary* 90 (1990): 18.

52. Susan George and Fabrizio Sabelli, ed., *Faith and Credit: the World Bank's Secular Empire* (Boulder, CO: Westview, 1994), p. 109.

53. Bjorn Lomborg, *The Skeptical Environmentalist* (New York: Cambridge University Press, 2001), pp. 280, 287, 291.

54. Ibid., pp. 303, 310, 320.

55. Norman Myers and Julian Simon, *Scarcity and Abundance: A Debate on the Environment* (New York: W. W. Norton, 1994), p. 65.

56. David Osborne and Ted Gaebler, *Reinventing Government* (Reading, MA: Addison-Wesley, 1992, originally 1990).

— 3 —
The Energy Crisis: Planning as Central Management

The oil crisis of 1973 stimulated extensive planning. Moreover, many of the plans were implemented. President Nixon addressed Congress on the issue four times in nine months, saying, "No single legislative area is more critical or more challenging to us as a people, however, than the subject of...the energy crisis."[1] At the time, the U.S. Geological Survey reported the country had only enough oil to last 10 years, enough natural gas to last 11 years, and enough uranium to last 13 years. President Carter called it "the moral equivalent of war." The Club of Rome projected 20 years of petroleum reserves.[2] The alarm continued. "The world is now lurching from one energy crisis to another, threatening at every turn to derail the global economy or disrupt its environmental support systems."[3] Even George W. Bush, a Republican who lauded the free market, spoke of an "energy crisis."

While the Club of Rome report exemplifies the predictive aspects of planning, the term is also used to mean central planning such as the Soviet five-year plans or Roosevelt's New Deal. An excellent example is mobilization for war. A nonmilitary example is the energy crisis, which stimulated great schemes of national and international planning as the industrial countries tried to counter the price increases and boycotts of the OPEC cartel. The Nixon administration moved immediately to centralize authority in the White House, instituted detailed control of prices and supply, and sought wartime authority. Although popularly known as the energy crisis, it was essentially an oil crisis. The centerpiece of American policy was supposed to be Project Energy Independence, a comprehensive program for regulating the economy, refining synthetic fuels, and radically changing consumer behavior.

MILITARY MOBILIZATION ANTECEDENTS

In November 1973, President Nixon ordered the Office of Defense and Civilian Mobilization to ration by fiat the quantities of petroleum to go to each state and to set the prices for crude oil at the wellhead. His authority was the Economic Stabilization Act, originally passed during World War II. The military framework of the Nixon program seemed appropriate. The proximate cause of the Arab boycott was their defeat by Israel in the Yom Kippur War a few weeks earlier. Nixon described the crisis in military terms, as President Carter did later. Military planning for both combat and logistics served as a model. In comparison to the planning of the Club of Rome, which was largely prediction with a little practical planning, the response to the energy crisis was largely planning with little prediction.

Nationwide mobilization can be traced back to Napoleon Bonaparte, who introduced total war and changed the scope of warfare from small mercenary to mass armies. Napoleon's humiliating defeat of the Prussian army stimulated it to modernize. In Berlin it established the Great General Staff of 30 officers to prepare plans, conduct war games, collect intelligence on potential enemies, and prepare maps. In 1810 the army established a War School. The plans were to be based on historical experience, the capabilities of artillery and other weapons, the terrain, transportation, and intelligence about potential enemies. War games, a true innovation, ranged from one based on a chess board with 1,666 squares to actual maneuvers in the field. For reality, one team of officers would compete against another, and the outcomes would be scored and analyzed. Rolling dice provided the element of chance. The general staff incorporated experience from the games and exercises into its plans. A practice mobilization in 1850, the first to use railways, proved a disaster. Many soldiers did not arrive at their battle locations for two months. This alerted the army to totally revise its plans.[4] For the remainder of the nineteenth century, the Prussian general staff was an out-standing success, contributing to quick victories in wars against Denmark, Austria, and France.

At the outbreak of World War I in 1914 German General Staff planning reached its acme in its invasion of Belgium and near defeat of France. Its heart was the rail transportation plan, ordering movement of more than three million men and 600,000 horses in 11,000 trains within 13 days.

During this time, 2,150 trains crossed the Rhine River bridge at Cologne at 10-minute intervals.[5] The plan was a triumph of bureaucratic management, put together by anonymous experts graded into a hierarchy with the kaiser at the head. But the authority of the kaiser was an illusion, for he was only a figurehead who could no longer control events once the mobilization began. When France blocked the German attack at the Battle of the Marne, the war degenerated into a horrific four-year stalemate with 10 million soldiers killed, 21 million wounded, and 8 million missing in action. Germany, Britain, and France mobilized their civilian economies to support the fighting. Great Britain established a ministry of munitions, placed the railways under government control, and imposed rationing. It controlled which jobs workers could take and set wages. The French government forced private companies into cartels, which it supervised. In Germany industry came under government control.

American mobilization for World War I was far more limited than the European mobilizations. During the first year of the war in Europe, Woodrow Wilson's administration did little to prepare the United States, on the assumption that preparation might encourage involvement. When the country did enter the war in 1917, the army and navy procured virtually all weapons and supplies by purchase from private companies. By the early twentieth century the economy was well integrated at the national level, which simplified planning and coordination. Giant corporations like U.S. Steel, Sears and Roebuck, and ATT were prominent, and trade associations were in place. The National Association of Manufacturers, the American Bankers Association, and the U.S. Chamber of Commerce occupied the peak, with numerous smaller and local associations around the country. Advocates pointed to the advantages of coordination and efficiency, while critics called them monopolies, cartels, and trusts.

Timeline for the Energy Crisis

1806	Prussian army organizes its Great General Staff for war planning and games.
1850	Prussian practice mobilization of army is a disaster.
1870	Prussia successfully mobilizes for war against France.
1914	Germany and Austria mobilize against France and Russia to begin World War I.
1917	United States mobilizes army and industry. War Industries Board established. Fuels Administration manages coal and oil.
1919	American Petroleum Institute established and fosters the Interstate Oil and Gas Compact.

1924	Herbert Hoover advocates industry codes for production quotas, wages, and prices.
1929	Pres. Hoover establishes the Committee on Social Trends chaired by William F. Ogburn.
1933	Congress passes the National Industrial Recovery Act. National Planning Board established.
1935	Supreme Court declares the NIRA unconstitutional. National Planning Board continues.
1939	Army and navy implement Industrial Mobilization Plan. Albert Einstein warns Roosevelt that an atomic bomb is possible.
1940	Roosevelt establishes the Office of Emergency Preparedness by Executive Order. Manhattan Project begins.
1941	Congress passes the War Powers Act. Harold Ickes appointed Petroleum Coordinator.
1943	Roosevelt establishes the Office of War Mobilization by executive order.
1944	Congress authorizes the Office of War Mobilization and Reconversion by law.
1945	Manhattan Project tests an atomic bomb and drops two on Japan.
1946	Congress passes the Atomic Energy Act. Atomic Energy Commission replaces the Manhattan Project.
1954	Atomic Energy Act amended to permit civilian reactors to generate electricity.
1959	Pres. Eisenhower establishes a quota for oil imports to protect domestic producers.
1973	OPEC raises oil prices four-fold, and Arabs embargo oil to United States and the Netherlands. Pres. Nixon establishes the Federal Energy Office and allocates supplies.
1974	Congress passes the Emergency Petroleum Allocation Act; crude oil price controlled.
1976	Project Independence report presented to Pres. Ford, who ignores it.
1977	Congress establishes the Department of Energy at Pres. Carter's request.
1978	Congress passes the National Energy Act at Pres. Carter's request.
1979	Iranians take 52 hostages at US embassy. OPEC doubles price of oil.
1981	Pres. Reagan abolishes price controls on crude oil. Terminates Synthetic Fuels Corp.
1986	Price of oil drops in half. Federal Energy Regulatory Commission administratively decontrols price of natural gas.
1991	Iraq defeated and expelled from Kuwait.
2003	United States invades Iraq.

President Wilson established the War Industries Board, headed by General Hugh Johnson. Its scope extended beyond army and navy procurement to include all industry. Its price-fixing committee, in spite of its grandiose title, depended on the cooperation of industry. For oligopolistic industries like steel, aluminum, and petroleum, company representatives would meet with the committee staff and soon reach consensus in most cases. Indeed, for the previous two decades they had been holding such meetings privately, and they now found the government imprimatur a benefit. They no longer had to worry about the Sherman Antitrust Act. In meetings of competitive industries, like textiles or lumber, however, agreement was often impossible to reach.[6] One method of pricing was "cost plus" (the cost to the manufacturer plus a reasonable profit). The corporations liked this because it guaranteed them their profit, but it undoubtedly was more expensive to the government. The producers had no incentive to compete in price, nor to keep their expenses low. The alternative was to negotiate between the board and the industry, which usually was cordial and generous to the manufacturers. In fact, price controls were only in effect for 12 to 18 months, so there was not much time to distort the market.

Fuel, food, and railroads were outside the purview of the War Industries Board. During the mobilization crisis in the winter of 1917–18, President Wilson officially seized the railroads. Earlier, Congress had established the Food Administration, headed by Herbert Hoover. The chief concern of the Fuels Administration was coal, especially during the winter. Oil was important to the British Royal Navy, and American wells supplied 85 percent of its demand. Mark Requa, who headed the oil division, had the producers cooperate to pool production, allocate supplies, and practice conservation.[7]

After the Armistice, the United States officially took a 12-year holiday from central planning. President Wilson had considered it only a temporary expedient and believed its continuation would give too much power to big business. The War Industries Board disbanded, along with the Munitions Board, and the food, fuel, and railroad administrations. The two Republican presidents who followed (Harding and Coolidge) did not believe that government should plan. Hoover, who took office in 1929, believed in some planning, but not as much as during the war. At the end of the war, Mark Requa of the Fuels Administration advocated loosening the antitrust laws to allow oil producers to establish quotas. When this came to naught, he left government to organize the private American Petroleum Institute, which eventually accomplished his goal.

PLANNING FOR THE NEW DEAL

Debate over the proper role of trade associations during the 1920s set the stage for official government planning in the 1930s. Herbert Hoover, by then secretary of commerce in President Warren G. Harding's cabinet, believed that cooperation via trade associations could increase efficiency, stabilize production, and better serve consumers. He strengthened his department's statistical data gathering even though publishing prices gave an advantage to oligopolistic firms. Hoover encouraged associations to simplify their products, to use standard accounting, and to advertise jointly. In 1924 the secretary announced in his annual report that "...now we must undertake nationwide elimination of waste."[8] The next year the Federal Trade Commission began to sponsor trade practice conferences in which industry representatives developed codes encompassing production quotas, wages, fair practices, and prices. Business trade associations would write their own rules, then submit them to the commission for approval and enforcement.[9]

In 1926 Hoover supported the recommendation of the Federal Oil Conservation Board to regulate drilling. Because crude oil flows naturally throughout its underground fields, one producer can capture more than its fair share by pumping as fast as possible. This led to many operators sinking wells and racing to capture their neighbor's oil. The rapid depletion wasted much oil. The remedy was to unitize the entire field by determining its natural limits and calculating quantities for each surface owner. Although intended to conserve a natural resource, a consequence was to limit production to the economic benefit of producers. Hoover considered this good because it would bring order to the market. In 1930 the Federal Trade Commission promoted a code for refined products like gasoline, extending regulation to refiners in California. This extended control from a natural resource (crude) to the refined product, thus going beyond the rationale of conservation. That year the Federal Oil Conservation Board began publishing statistics useful in coordinating between the oil-producing states and urged them to join together through an interstate compact to apportion state quotas. Both as secretary of commerce and as president, Hoover emphasized industrial cooperation and de-emphasized antitrust laws. Although petroleum was not the only arena subject to government-sponsored codes, it was more prominent. Other codes covered were for construction, metals, textiles, and furniture.

The codes Hoover promoted as commerce secretary and later as president did not derive from a central government desire to bring industry under control, but from the desires of business leaders to create a more rational environment for themselves. Unlike Europe, the origins of the codes were not from the left (socialism) but from the right.

The Great Depression and the election of Roosevelt put a new twist on the industrial codes of the 1920s, culminating in the National Industrial

Recovery Act of 1933. Roosevelt himself was an experimenter who would try a variety of programs until he found one that worked. With the devastation of the Depression, businessmen came to welcome experimentation as well. In 1931 the U.S. Chamber of Commerce recommended controlling production, unemployment insurance, and a shorter workweek. Gerard Swope, president of the General Electric Company, announced his famous plan to regulate prices and production and to establish unemployment insurance, workman's compensation, old-age insurance, and employee representation. In March 1933 the Chamber of Commerce recommended work sharing and minimum wage rates. Furthermore it recommended establishment of a government agency to supervise this program. Thus the new administration operated in a proplanning context inherited from the previous two Republican administrations.

Only a few of Roosevelt's advisors advocated planning from other than a capitalistic perspective. One of them, Rexford Tugwell, wrote that planning should be universal and comprehensive. Its goal, under the guidance of experts, was to serve the public first, not to enhance profits for business. George Galloway wrote that "industrial planning is the conscious guidance of our industrial life by collective methods."[10] When Congress debated the NIRA bill, opponents said it too closely resembled Mussolini's edicts and German cartels.[11]

The chief embodiment of government planning, whether derived from the left or the right, was the Recovery Act, passed by Congress and signed by the president in June 1933, one of the highlights of Roosevelt's first hundred days. The author of the first draft of a bill was Hugh Johnson, the World War I general who headed the War Industries Board. To stimulate the economy by direct spending, the law included a large public works portion to be administered separately.

Roosevelt appointed General Johnson to head the National Recovery Administration established in the act. The heart of the NRA was the research and planning division, which was responsible for analyzing and approving codes for the various industries and more generally the balance among industries, wage policies, and capital goods. Lack of statistics was a great problem, since companies tried to keep their data secret. Even when statistics were available, NRA administrators from outside the research and planning division often ignored them and made decisions according to preconceived ideas.[12]

Section 9c of the Recovery Act specifically established nationwide regulation of oil production, and the American Petroleum Institute took the lead in organizing the code that combined price fixing and production and inventory control.[13] Although the act supposedly gave top priority to developing a code for coal mining, this proved difficult due to the bitter labor disputes in the preceding years.

When the Supreme Court declared the National Industrial Recovery Act unconstitutional less than two years after its passage, the U.S. experiment with central planning was over. Roosevelt and top officials accepted this, and many seemed to breathe a sigh of relief. Problems with the Recovery Act were readily apparent by this time. Writing only two years after the Supreme Court ruling, Charles Roos, formerly the chief of the NRA Division of Analysis and Research, described the confusion and contradictions. He noted that the bill was sent to Congress hastily and passed quickly. Roos warned future planners to consider the ramifications fully and to get the widest possible debate, "lest the 'cure' itself transcend the pain of the disease."[14] He concluded that the act lacked definite policies and objectives and that it embodied contradictory goals (such as both shorter hours and higher incomes). The method of code writing was haphazard, consisting of bargaining between industry and labor, without concern with their relative strengths in different industries or concern for the consumer, who had to pay higher prices. On the plus side of Roos' ledger, the NRA was a giant experiment, teaching lessons about the ramifications of one decision on other factors, for example, that controlling one price soon leads to the need to control others, including discounts, interest on unpaid balances, and so forth.

The replacement for the comprehensive Recovery Act was a series of individual programs such as the Works Progress Administration. The most important ones for electricity were the Tennessee Valley Authority and the Bonneville Power Administration. Ironically, constructing all the dams on the Columbia River oversupplied it with electricity. In the 1930s the Northwest was not very industrialized, but when World War II came, its overcapacity was used up by new factories for aircraft and weapons. Aluminum refining demanded huge quantities. The secret project to build the atomic bomb established a plutonium plant at Hanford, Washington. The other big plant was in Oak Ridge, Tennessee, to take advantage of TVA electricity.

Upon the Supreme Court declaration that the Recovery Act was unconstitutional, the oil industry, under the leadership of the American Petroleum Institute, found an alternative in the Interstate Oil and Gas Compact, which was a treaty among the six largest producing states to coordinate their quotas. Congress voted to approve the compact, and the interstate cartel, with headquarters in Oklahoma City, operated with little change from 1935 until the 1973 oil crisis, at which time Congress took back authority for regulating production. The Compact Commission continues to advocate for higher prices for producers. In 1999 it published a report titled *A Battle for Survival? The Real Story Behind Low Oil Prices,* which described the "dramatic impact of prolonged low oil prices on the United States" and "paints a picture of an industry in trouble, a growing threat to national security, and the loss of income to individual royalty owners, states and the federal government."[15]

One division of the NRA that continued until 1943 was the National Planning Board (renamed the National Resources Board for its second year of existence). When the Supreme Court struck down the Recovery Act, it became the National Planning Committee and then, in 1939, became the National Resources Planning Board. In spite of its four names and precarious legal status, its membership, staff, and goals remained fairly stable. Its purpose was to plan and to aid the planning of other agencies. Within the NRA, its location was on the public works side, not the industrial code side. The board's greatest success was in planning for water. In public works, the Army Corps of Engineers often thwarted its proposals. The leaders of the board came from the field of city planning. The chairman for all 10 years was Frederic Delano, the head of the National Capital Park and Planning Commission for Washington, DC. Delano had formerly developed plans in Chicago and New York and was the president's uncle. He brought his staff director, Charles Eliot, from the city to the national board as executive director. Charles E. Merriam, a political scientist from the University of Chicago, was the most influential board member. His background and interests were in social policies rather than water and land.[16]

The Resources Board devoted a great deal of effort to the planning process itself, seeking to foster planning in other federal agencies and in the states and citiesft published 107 reports on that topic, far more than on any other one. Because many problems extended beyond the boundaries of a single state, the board promoted regional planning, then a new concept. Many of the regions were river basins like the Columbia or the Tennessee. Delano and Eliot had their backgrounds in city planning and encouraged more of it. Many New Deal grant programs required the city to demonstrate how a project would fit their city's plan, thereby expanding the process. Marion Clawson, in weighing the influence of the board, gave it high marks for stimulating and developing ideas, but lower marks for its general planning and coordinating function. He considered it best in water planning, less good for land, mineral, or energy, weak for public works, and nonexistent for agriculture. Clawson noted with regret that the board had only a minor role in war planning, even though the effort was in bad shape. A number of board alumni did move to important positions in wartime agencies, however.[17]

Although the Resources Board, under its four names, survived a decade and was a small positive influence, it never achieved the comprehensive control of federal agencies for which Roosevelt had hoped. In 1936 he took another tack by appointing Louis Brownlow to head an ad hoc committee on administrative management to recommend how to coordinate. The next year it reported that "the time has come to set our house in order" by consolidating agencies, giving the president six administrative assistants in the White House and establishing a "clearinghouse" planning agency. Brownlow himself had said "we cannot escape planning." Roosevelt, however, avoided the term because of conservative opposition and preferred to talk about "management." Roosevelt did like the idea of consolidating agencies and submitted a draft bill to Congress to expand his authority to reorganize the government. Unfortunately for him, the bill got tied up in his proposal to reorganize the Supreme Court by packing it with new appointees who would favor the New Deal. Public opposition to packing the court was intense and spilled over to opposition to the reorganization bill. Moreover, many interest groups opposed giving the president authority to reorganize agencies. The American Legion testified against moving the Veterans Administration into a proposed department of welfare. Social workers opposed moving the Children's Bureau from labor to welfare. Physicians feared the socialistic impulses for adding the Public Health Service to such a department.[18]

When Congress addressed the problems of the coal industry in passing the Bituminous Coal Act in 1937, it took the perspective of the Great Depression. The law's goals were to fix minimum prices, to prevent opening too many mines, and to limit excess capacity. The coal division of the Department of the Interior was supposed to determine prices according to a full valuation of mine property, a task that would have taken years to complete. The law provided that government should fix prices high enough to guarantee profitability to all producers, irrespective of their costs.[19]

MOBILIZATION FOR WORLD WAR II

This mobilization had many false starts. In 1939, just before the German invasion of Poland, the military activated its Industrial Mobilization Plan, but it was abandoned as too warlike for a country officially neutral.

Furthermore Roosevelt did not like not being in control personally. In May 1940 he issued an executive order creating the Office of Emergency Preparedness and then reached back to a World War I law to activate the National Defense Advisory Commission. He thereby avoided asking Congress for new legislation at a time when it still favored neutrality, and when public opinion did not favor stronger steps. Roosevelt did not want to cause himself problems in the presidential election that November.[20] The United States did not move onto a wartime footing until the middle of 1941, a full year after the defeat of France, and many months after the Battle of Britain began.

The National Defense Advisory Commission was to prove to be another false start, for within a few months, by executive order, Roosevelt moved its top leaders and many of its functions to the new Office of Production Management. Other agencies competing for resources covered Lend Lease, Selective Service, and ones for transportation, rubber, labor, and prices. The system was ad hoc, and control remained with the president personally. As the agencies competed for money and priority, the quarreling of their leaders reached such intensity in the spring of 1943 that the episode earned the nickname "the Battle of Washington."[21] From Capitol Hill, Senator Harry Truman, who chaired a special committee investigating the defense program, blasted the ad hoc system as mismanaged and undermining the war effort.

In May 1943 President Roosevelt established the Office of War Mobilization, with broad authority, and named James Byrnes as its director. Byrnes had served as a representative and senator from South Carolina for 28 years, was a loyal Democrat, and was a devoted New Dealer. Byrnes viewed his job as being the president's chief assistant, operating with a personal link to him. He resisted building up the staff of the OWM, preferring to save his participation for deciding only the most important issues. The small staff of 16 definitely did not plan, but rather identified problems and arranged for Byrnes to arbitrate and arrange compromises. His role was personal.

Almost as soon as it was created, the OWM recognized the future need for reconversion at the end of the war. Most people considered the transition at the end of World War I bad. Congress passed legislation establishing the Office of War Mobilization and Reconversion in October 1944. Roosevelt and Byrnes actually had preferred to do this by executive order, like the OWM a year and a half earlier. It would have given the president more control and flexibility, but Congress was determined to legislate, and it seemed best to cooperate. The law directed the OWMR to formulate plans and develop procedures, thus giving planning a larger role than under the OWM arrangement. One of the effects was to increase the staff to a high of 146, with the addition of statisticians, economists, and lawyers.[22]

Government policy regarding energy was part of its overall policy on industrial production. The different fuels had their own agencies such as the Petroleum Administration for War, headed by Harold Ickes, and the Solid Fuels Administration. Due to higher demands for oil, and to the German submarines sinking of tanker ships along the East Coast, Ickes built a pipeline from Texas to the Northeast. The petroleum industry opposed this because it feared that, after the war, the pipeline would become a government-owned competitor. Eventually the oil industry got its way, and the pipeline was converted to natural gas.

The Manhattan Project to build the atomic bomb was another legacy of World War II that has shaped governmental planning on energy to the present day. Its importance is not that it was well done itself, but that its dramatic success became an ideal for postwar organization of major projects. Americans came to believe that when the government devoted enough money and personnel, it could quickly solve nearly impossible scientific and technical problems. The slogan soon became, "If we can build the A bomb, surely we can solve the problem of _____." In the 1960s this faith in government technology got another giant boost from the NASA success in landing a man on the moon.

A second belief was that a large government bureaucracy was the best way to achieve the goal. The Manhattan Project employed thousands of workers in newly constructed plants in Oak Ridge, Tennessee; Hanford, Washington; Los Alamos, New Mexico; and a dozen other sites. The project's accomplishment cannot be attributed to good planning. Indeed, blunders and inefficiency were characteristics. It wasted money and coordinated poorly. The scientists considered the commanding engineer general, Leslie Groves, pompous and scientifically ignorant. Nevertheless it achieved its goal of developing a totally new weapon. Moreover, it achieved this in absolute secrecy.

After the war ended, Congress converted the military Manhattan Project into the civilian Atomic Energy Commission (AEC). The new agency continued many of the military aspects by insisting on secrecy and operating by command. It gained a reputation for arrogance and disregard for environmental protection. In spite of the emergence of civilian electric generating reactors owned by private utilities, the commission continued its old ways. This situation changed about 1970. One factor was that knowledge of nuclear technology became more widespread outside the AEC. By then a number of scientists, chiefly in universities, had the expertise to challenge the government experts. In addition, Congress passed and President Nixon signed the National Environmental Policy Act, which provided that when the AEC issued a license for a civilian electric generating plant, it also had to prepare an environmental impact statement. At first the AEC did not believe that the law applied to it, but in 1971 the Circuit Court of Appeals ruled that the commission did have to comply in the case of the Calvert Cliffs plant of the Baltimore Gas and Electric Company. Things were never the same, as environmentalists began to intervene frequently. The power of secrecy slipped away.

Besides the Office of War Mobilization and the Manhattan Project, World War II left a third legacy for planning: operations research. This is the application of mathematics, science, and technology to solve problems of management. A key to solving a problem is to use a multidisciplinary team. Operations research seeks to use personnel, equipment, and supplies most efficiently. In 1937 the British Royal Air Force, having secretly developed radar, sought the most effective way to use it to locate enemy aircraft, soon recognizing the need to combine the skills of engineers, mathematicians, surveyors, and fighter pilots. In 1942 a British team visited the United States, where it introduced operations research. By the end of the war the U.S. Air Force had 26 teams.

At the end of the war, operations researchers moved from the military to industry. One U.S. Air Force officer who later rose to prominence was Robert S. McNamara. Trained in finance and statistics at Harvard Business School, he was given an army commission. Captain McNamara went to England to straighten out problems with the B-17 bomber. Soon promoted to major, he went to India and Burma to coordinate war supplies flown across the Himalayas to China. Four months later he was a lieutenant colonel assigned to the Pentagon, and later to the Air Technical Command in Ohio. After the war, McNamara's commander, Colonel Tex Thornton, assembled a team of a dozen of his management experts to look collectively for the corporation that would offer them the best employment deal. The winner was the Ford Motor Company, where over the next 20 years McNamara and two others rose to become presidents and six became vice presidents.[23] In 1961 President Kennedy appointed McNamara to be secretary of defense, where he revolutionized management with PPBS—the Program Planning and Budgeting System. During the Vietnam War, he applied operations research by keeping track of the body count of dead Viet Cong soldiers.

By 1970 operations research appeared to be the ideal way for industry to deal with resource problems. The analytical power of mainframe computers was added to the essentials of systems orientation, interdisciplinary teams, and the scientific method. (In making his predications for the Club of Rome report, Jay Forrester's analysis was systematic and scientific, but not interdisciplinary.) The methodology begins with formulating the problem in terms of statistical variables. This requires detailed and complete understanding of the situation. The variables must be quantifiable; if it cannot be quantified, it cannot be analyzed. The researcher develops a model and prepares a flow chart. The next step is to determine risk and uncertainty. Risk means alternatives where the probabilities are known, and uncertainty means where they are unknown. Uncertainty can be a huge stumbling block.

THE 1973 CRISIS

In terms of prediction as an aspect of planning, the national government did diagnose, to a minor extent, the coming shortage of oil. The oil division of the Department of the Interior, which tracked the exploration, development, and production of crude oil, recognized that the country was using more oil than it was discovering. This pattern extended back for decades. In 1947 the United States became a net importer, then getting most of its imports from Venezuela. Imports were cheap and seemingly limitless, which domestic producers considered a threat to their profits. In 1959 they persuaded the Eisenhower administration to restrict imports to 12 percent, thereby supporting the domestic price. One of the arguments advanced in favor of the program was that it would protect the country in case of war. By 1971 supplies were very tight. Natural gas, which the government had tightly controlled since 1938, was scarce too. Conflicts between industry and environmentalists emerged, especially about drilling off shore. In 1969 a well six miles off the coast of Santa Barbara exploded, releasing 235,000 gallons that spread into a slick covering 800 square miles.

The Interior Department asked its public-private advisory body, the National Petroleum Council, to make a comprehensive assessment. The department further asked the council to include coal and electricity, because of the interaction between petroleum and other fuels, and to title the report *U.S. Energy Outlook.*[24] Of 30 members, only one was a government official. The other 29 were drawn from petroleum corporations plus the American Petroleum Institute. No consumers were represented. The council used a series of subcommittees staffed by experts to do the actual analysis. A single Interior Department official served as a cochair of each subcommittee. When issued in March 1973, the report had contradictory conclusions. Its statistics showed that a shortage was coming soon, but its recommendations were that policies should continue the same, especially the import quota. These were the conclusions of the industry representatives who dominated the council.

Meanwhile in the White House, several similar predictions emerged about a looming oil shortage. In the spring of 1972 General George A. Lincoln, head of the Office of Emergency Planning, had foreseen the danger. No one paid much attention, and he soon retired and his office was abolished. Next a State Department expert on the Middle East came to the White House staff, where he warned of an oil shortage and recommended better treatment of the Arabs. He was sent back to Foggy Bottom. His successor was Charles DiBona, a systems analyst, whose statistical analysis predicted a shortage. By the end of the summer Nixon decided he needed a politician, so he appointed Governor John Love of Colorado, whom the president described as an "energy czar." Love lasted until December. Nixon's management of the energy crisis as it unfolded over the next year had roots in the New Deal and the World War II mobilization.

Central management is the other aspect of planning besides prediction. This had been a centerpiece of the New Deal and is most famous for its socialist and totalitarian incarnations: the Soviet five-year plans, Mussolini's corporativism, the Nazi's four-year plans, and the British Labour government's nationalization of coal, electricity, and other industries after 1945. In one of the supreme ironies of history, the Republican Richard Nixon embraced elements of central planning for dealing with the oil crisis.

Throughout the nineteenth century and up to 1929, few Americans imagined central planning as an alternative to the free market system. First, no government had the information and power to accomplish central planning. Second, no foreign models existed until the Soviet and Italian initiatives in the 1920s, and these did not appear very successful. Third, until the stock market crash, the market system appeared very successful. Yet the American system was not entirely a free market. By the turn of the century, monopolies and cartels dominated many industrial sectors including oil and electricity. In terms of their choice of political party, the early twentieth-century critics of the free market tended to be Republicans such as Theodore Roosevelt and even Herbert Hoover. On balance, however, the GOP was firmly on the side of business.

The Democrats were still mired in the Southern defeat in the Civil War, and many in the north were gold bugs, that is, conservative proponents of the gold standard. The party's northern component of immigrants and their children loyal to the city machines such as in New York and Boston were not numerous enough to win a national election. Franklin Roosevelt reversed the Democratic fortunes in 1932 by forging a coalition of the northern urban voters with the Solid South. But voters in this combination were not particularly opposed to the market system. It was the Great Depression that they opposed. Much of the impetus for central planning derived from FDR's personality. He was a doer who wanted to help people, to plant trees, and to dam rivers. He was psychologically suited to action, as demonstrated in his jobs as a state senator, assistant secretary of the navy, and governor of New York. Roosevelt's proclivity for central management derived from personal characteristics, not adherence to an ideology. Indeed he was a rather conventional man, loyal to his family, his farm, his Hyde Park community, his church, and his political party. He inherited his Democratic affiliation from his father, who was a bit of a gold bug. FDR was hardly a leftist.

While Roosevelt was ideologically conventional, many of his close advisors were fervent liberals. For the most part, he got his ideology second hand. Tugwell, who wrote campaign speeches in 1932, was strongly liberal. Although FDR campaigned that fall as a fiscal conservative, promising to cut the federal budget, before his March 4 inauguration he was persuaded of the benefits of government spending to stimulate the economy, an idea he got indirectly from John Maynard Keynes, the British economist. Other aspects of FDR's energy policies, such as the Public Utilities Holding Company Act of 1935 and the Natural Gas Act of 1938, reflected his desire to break the power of monopolies. Roosevelt and the New Dealers believed that private companies were abusing their monopoly power by charging high prices and denying service to many northern cities. Before too many years passed, Roosevelt became a bitter enemy, calling them "economic royalists" and "money changers in the Temple."

The central planning for the energy crisis had a further root in the Roosevelt administration, in this case the World War II mobilization. The legislative basis was the War Powers Act passed a few days after the attack on Pearl Harbor and expanded extensively four months later. The law gave the president extreme authority to set up whatever agencies and regulations he needed to manage the war, without further congressional action.[25] The law gave the president authority to set prices, determine quantity, and ration supplies. After the end of World War II the law remained on the books, but was not used. The issue of the law surfaced again in 1970 when President Nixon faced the problem of inflation. Democrats in Congress criticized him by saying that he had the authority under the War Powers Act to freeze prices. In response to Nixon's claim that a 30-year-old wartime law was not the appropriate legal basis, the Democratic majority in Congress amended the law and renamed it the Economic Stabilization Act. Few political observers at the time considered this to be more than partisan posturing. They could not conceive of a Republican president actually controlling the economy by fiat. Economists considered the possibility ludicrous.

Nixon fooled them. In 1971 inflation had risen to the unprecedented level of 4 percent. In a stunning move on August 15, the president declared that under the authority of the Economic Stabilization Act he was freezing all prices. The plan proved a major failure and within a year was abandoned. The Nixon administration removed controls over the economy with the exception of two industries: petroleum and construction. Nixon considered these to be too important to exempt entirely. Central management was to continue, although the hand of government was to be light. Regulations were flexible enough to permit prices to rise and fall to the market level determined by supply and demand. When the oil boycott occurred, the control system was in place.

American policy regarding oil paid almost no attention to international aspects until about 1971. After the Supreme Court ruled the National Industrial Recovery Act unconstitutional in 1935, policy barely occurred at the national level, but was coordinated by the oil-producing states through the Interstate Oil and Gas Compact. The huge surplus Texas could produce gave it power to increase or decrease its quantity, thereby determining the national supply, and with it price. During the 1960s the United States quietly increased its dependence on imports. The international price depended on the import quota, which depended on the compact and the state of Texas. The domestic price was $3 a barrel. The actual cost of imports was $1.50 a barrel, but the quota raised it to the domestic price.

Prices for foreign oil were set by negotiations among the seven major petroleum companies and the exporters. The majors, known as the Seven Sisters, were Standard Oil of New Jersey (later Exxon), Mobil, Shell, Standard Oil of Indiana (later Amoco), Texaco, Gulf, and Standard of California (Chevron). In effect they dictated prices, acting like a cartel. This changed about 1970 as demand increased and American domestic supplies could not keep pace. For a long time, the companies had been hostile to help from the government, and this attitude continued until 1973.

The first threat to this system from a foreign government came in 1969 when the radical Colonel Muammar Gaddafi overthrew the pro-Western king of Libya. Libya did not export much oil to the United States, but it did export to Europe. Moreover, its crude was high quality and cheap to transport to Europe. The company active in Libya was not one of the seven majors, but a small independent company: Occidental. Gaddafi demanded Occidental raise its payments by 20 percent, and when it refused, he forced it to cut its production by over half. The Seven Sisters did nothing to help their rival. Neither did they ask the U.S. government to assist them. The companies simply did not consider it to be a suitable subject for governmental involvement. Despite Gaddafi's anti-Western stance, the United States chose to do nothing. Its European allies chose to curry favor with the colonel. France sold him 100 advanced jet aircraft (which were later used against Israel). West Germany in particular ingratiated itself with Libya. Secretary of State Henry Kissinger believed that this toadying sent a signal to other oil producers that the Americans and Europeans would not protect their friends.[26]

In 1971, for the first time all OPEC countries decided to bargain with the oil companies as a unit. Although the U.S. Department of State did offer to help and sent its undersecretary on a trip to the Middle East, the companies rebuffed the offer. The consequence was that the exporters raised their prices by 40 cents a barrel, about 25 percent. This was still cheaper than the domestic price. The following year OPEC turned up the pressure. It raised prices a further 9 percent. More important, under the term "participation," it asked that the Seven Sisters turn over 20 percent ownership of their installations to the exporters and proposed that eventually ownership reach 51 percent, a controlling interest. The companies refused the American government's offer to help negotiate new terms and compensation with OPEC.

Nixon anticipated the looming oil problems. In June 1971 he had sent Congress a message urging expansion of alternative energy sources, observing proudly that it was the first time a president had ever sent a comprehensive energy proposal. Little came of it, however. In April 1973 Nixon sent a major energy proposal to Congress with initiatives to conserve energy, expand domestic production, and reorganize government agencies. He also replaced the oil import quota with a simpler system based on fees.

The oil boycott that disrupted the American and world economies that fall had three causes. First was the long-term depletion of reserves described by the National Petroleum Council report. New domestic discoveries were falling behind consumption. The American demand for additional petroleum could only be met by imports. The second cause was the OPEC cartel. OPEC was established in 1960 under Venezuelan leadership to seek higher profits and the next year first tried to use its monopoly power. That attempt failed because it did not control a large enough proportion of the world supply, but by 1973 it was successful. OPEC doubled the price immediately and over the next few months continued to increase it to six times the original level. The third cause was the hatred of Israel by the seven Arab countries at the core of OPEC: Saudi Arabia, Kuwait, Iraq, Abu Dhabi, Qatar, Libya, and Algeria. As gesture of support for their comrades in Egypt and Syria, who had just lost the Yom Kippur War to Israel, they refused to export oil to countries that had supported the Jewish State. France, Japan, and Italy quickly apologized, so in the end only the United States and the Netherlands were embargoed. Because crude oil is nearly the same everywhere, these two countries eventually made up their shortfalls from non-Arab sources, but for several months during the winter, imports to the United States were down by 10 percent.

The public reaction came close to panic. The chief manifestation was not higher prices at the gasoline pump, although they did go up. Motorists wanted to avoid running out so they filled up their tanks frequently, causing long lines. Many service station operators voluntarily imposed a limit of five gallons, which made the gas lines longer. The state of Oregon decreed that cars with even-numbered license plates could buy only on the even-numbered days of the week, and odd-numbered plates on odd days. Soon other states copied the Oregon plan. In January truckers who were short of diesel fuel declared a strike and blockaded highways. Several truckers who did not honor the strike were killed by snipers, and governors of a dozen states mobilized the National Guard. President Nixon intervened using authority under the Economic Stabilization Act (originally the War Powers Act). Each state was to receive an allocation based on its volume the previous August, a date chosen because it was the most recent for which data was available. The problem was that those figures were for the summer pattern of high consumption of gasoline and low consumption of heating oil, the exact opposite of the winter demand pattern.

Two months after the embargo began, Nixon established the temporary Federal Energy Office in the executive office of the president and named William Simon to be the energy czar. Simon was a banker who was the deputy secretary of the Treasury. Two days after making the appointment, Nixon discussed it in a cabinet meeting, comparing it favorably to Hitler's appointment of Albert Speer to direct the Nazi armament and munitions in the midst of World War II.[27] The new czar suppressed his capitalistic instincts and managed the Federal Energy Office according to command and control. As of January 1974, its centerpiece was mandatory allocation of gasoline and oil based on usage according the most recent data available, which skewed production away from heating oil. Furthermore the system could not adjust to regional demands, so some locations had adequate supplies and others had shortages. Simon later wrote that "as for the centralized allocation process itself, the kindest thing I can say about it is that it was a disaster."[28] When he tried to break the bottleneck of gasoline, his civil service staff refused to cooperate, saying that it was reckless. Consequently he and a few political aides drew up all of the paperwork themselves, and in his words, "by the end of a 36 hour period, the gas lines were cracked."[29]

In April Simon moved up to become secretary of the treasury, and Nixon replaced him with John Sawhill, an economist.

A greater question than Simon's apostasy is why Nixon himself decided to control energy supply by command and control. He had personal experience in government control of the economy as a young lawyer when he worked for the wartime Office of Price Administration in 1942, before entering the navy. He wrote that "my personal experience at the OPA had convinced me that rationing does not work well even in wartime when patriotism inspires sacrifice. I knew that in peacetime an enormous black market would develop and the entire program would become a fiasco.... I was sure rationing would end up being a cure worse than the illness."[30] Nevertheless he went so far as to order ration coupons printed, although they were never issued.

Congress, under Democratic control in both houses, pressed Nixon for a strong governmental role. Senator Henry Jackson, chairman of the Interior Committee, led passage of an emergency energy bill, which gave the president broad power to ration gasoline, conserve oil, and reduce environmental safeguards. The bill would roll back the price of domestic crude oil to $5.25 per barrel, which caused Nixon to veto it. Congress did pass it, and the president signed a number of other laws. The Federal Energy Administration Act made the temporary office permanent. The Emergency Petroleum Allocation Act directed the president to establish mandatory allocation of crude oil, fuel oil, and gasoline within 15 days. The program was to set the quantities and prices. In an exercise of double speak, Section 4 (b) (1) (I) directed that the program provide for the "minimization of economic distortion, inflexibility and unnecessary interference with market mechanisms." Section 8 set the maximum price of crude oil at $7.66 per barrel, a provision that cursed the program for its entire existence. Congress and the president apparently did not understand the difference between average price and marginal price. The cost to bring in new production was then about $10 per barrel, so an oil company had no incentive to do so. Indeed if a well had costs above the legal limit of $7.66, the company tried to find an excuse to take it off stream. Within a few years the Federal Energy Administration and its successor, the Department of Energy, developed a regulatory monster of at least 17 different prices for crude.

When Nixon established the Federal Energy Office in December 1973, he also announced Project Independence, which was supposed to make the United States independent of foreign imports within 12 years. The country would refine oil from shale deposits in Colorado and Wyoming. It would make synthetic gasoline from coal in West Virginia and North Dakota.

It would pump out more oil from abandoned wells and exploit the reserves in Alaska. Project Independence would be comprehensive, expanding use of coal, natural gas, and nuclear power. Moreover, by 1985 the United States would have so much oil that it could export it to Japan, so the Asian ally would not be dependent on OPEC. The White House named a blue ribbon panel and backed them up with dozens of experts.

Nixon and Kissinger recognized the importance of cooperating with Europe and Japan, because those countries imported 16 million barrels per day, compared to 6 million for the United States.[31] Moreover in percentage terms, they depended far more on imports than the United States: 75 compared to 39 percent. Oil was the same worldwide, so reducing demand one place would make more available for others. Nixon called the Washington Energy Conference in February 1974, where the industrial countries established the International Energy Agency. The American goal was to unify the response. In the immediate aftermath of the crisis, France had tried to negotiate special deals with the Arabs, and Britain, Germany, and Italy had been inclined that way, but Nixon and Kissinger were able to persuade the Europeans to stick together, in part by tying energy policy to the military defense of Europe.

In 1974 Richard Nixon had a greater problem than energy. The Watergate scandal closed in and forced him to resign on August 9. When the Project Independence task force sent its preliminary draft report to the White House in the fall, the new president, Jerry Ford, was appalled. The draft was pessimistic about the possibility of becoming independent of oil imports without a massive conservation effort. Furthermore, Ford, whose entire career was spent as a congressman from Grand Rapids, Michigan, had a midwestern, conservative Republican aversion to grandiose New Deal government corporations. Ford fired his FEA head, Sawhill, and replaced him with Frank G. Zarb, making him the seventh energy czar. The ambitious Project Independence withered away.

While the new president was averse to Project Independence due to his free enterprise background, within a few months two of his top advisors persuaded him otherwise. The chairman of the Republican Party, Rogers Morton, worried that Ford's lack of action would lead to defeat in the 1976 election. The new vice president, Nelson Rockefeller, agreed about the political danger, and also believed that the free market would not solve the energy problem.[32] Rockefeller, who was the former governor of New York and the grandson of the founder of Standard Oil Company, feared for its monopoly power 60 years earlier. Ford had named him to the position on the basis of his 15 years as a liberal governor who could attract support from liberals and moderates. Back in 1940, despite the fact that he was a Republican, Roosevelt had named him the coordinator of inter-American affairs and later to the position of assistant secretary of state for the American republics. Later Truman named him head of the International Development Advisory Board. Still later he served in the Republican administration of Dwight Eisenhower, before winning the first of four elections for governor. Unlike most members of the GOP, Rockefeller believed in big government. Once appointed vice president, he began planning for the Energy Independence Authority, which was to be an agency funded with $20 billion in equity and $100 billion in lending authority. It was modeled on the Reconstruction Finance Corporation of the New Deal. Projects might include plants to manufacture synthetic fuels from coal and shale, a trans-Canada oil pipeline, and a uranium enrichment plant. Loans would be subsidized.

Ford supported the authority reluctantly, persuaded only by its political symbolism. Several key members of his administration opposed the concept because it violated the principle of the free market, would crowd out other investments, would become a pork barrel for Democrats who controlled Congress, and could never be abolished. Ford announced the authority proposal in a speech to the AFL-CIO Construction Trades Union convention in September 1975, stressing that it would create jobs and reduce oil imports. But it was not in tune with the Ford administration themes, and as the election campaign heated up, this second incarnation of Project Independence faded away.

Project Independence had also caused problems with American allies in Europe and Japan, who depended far more on oil imports than did the United States. Even with the North Sea fields coming on stream, Europe produced only about half of its consumption. Japan had no domestic sources and recognized that the promise to export synthetic fuels by 1985 was phony. Kissinger immediately started calling it Project *Inter*dependence. He also began dropping hints that the United States might invade one or two OPEC members to seize their fields. He spoke of "political weapons," "conflict," and "war" in speeches at the United Nations and the World Energy Conference. He also talked about using food as a weapon to trade for oil, insinuating the threat of starvation for OPEC members.

The Strategic Petroleum Reserve was one of the less controversial aspects of Ford's years as president. The government was to purchase and store 500 million barrels of crude oil as a cushion against another supply disruption. Several existing caverns in salt domes in Louisiana and Texas were good sites because they were cheap and close to existing pipelines and refineries. Oil was pumped down into the caverns, where it would be ready for a future emergency. Although authorized by law in 1975, no oil was pumped in for two years.

CARTER'S ENERGY PLAN

The 1976 presidential campaign concentrated on energy as one of the top issues. President Ford had followed an inconsistent policy, on one hand rejecting government allocation and Project Independence, but on the other hand, in the fall of 1975, signing the Energy Policy and Conservation Act, which tightened regulation of crude oil, mandated automobile mileage standards, and required each state to have a conservation plan. His opponent, Governor Jimmy Carter, promised, if elected, to have a comprehensive energy plan announced within 90 days of his inauguration. True to his word, after his election victory he set up a task force that gathered data, consulted experts, and on April 20 issued a report titled *The National Energy Plan*.[33] The plan proposed little to increase the supply of oil, but a lot to reduce consumption. The government would use taxes and direct regulation to continue to control the prices for crude oil, but would gradually increase the level to meet the world market price. The government would raise the tax on gasoline, tax windfall profits on crude oil, and force electric utilities to switch from natural gas to coal. Imports would be reduced from 16 million barrels a day to 6 million barrels. Contemporaneously Carter proposed legislation establishing the Department of Energy. Its core was the old Atomic Energy Commission plus the Federal Energy Administration from the executive office of the president and the independent Federal Power Commission. Congress enacted the bill with little debate, and Carter signed it.

The next step was to persuade Congress to enact the recommendations contained in the national energy plan. Supposedly this would give the United States a coherent and comprehensive energy policy. Carter submitted a massive draft bill covering oil, natural gas, nuclear energy, and coal. It would subsidize solar and wind power and force conservation. When actually enacted in October 1978, it did only a few of these things. Congress decided that the voters did not want higher taxes on gasoline, and that oil producers did not want a tax on windfall profits. Indeed the producers did not consider their profits to be windfalls.

Within two months, fate turned further against President Carter, when oil workers in Iran struck, stopping a supply of 5.5 million barrels a day to the United States, Europe, and Japan. The strike was the first phase of a conservative Islamic revolution that toppled the pro-Western government of the Shah. His replacement was an exiled religious leader, Ayatollah Ruhollah

Khomeini. The shortfall of Iranian oil sent the price up immediately. OPEC soon moved to take advantage of the shortage by raising the price to $14, then to $20, then to $28 by the end of the year. Looking back to 1973 the real price (adjusted for inflation) had quadrupled in 1973–74, remained at a plateau for five years, then doubled in 1979–80. The price rise initiated a recession. The other disaster for Carter was that the Iranian revolutionaries, in violation of international law, had entered the American Embassy in Tehran and arrested 52 diplomats stationed there. Khomeini held them hostage and taunted the president. The recession and the hostage situation were major factors in Carter's defeat in the 1980 election.

Back home, Carter revived the Ford-Rockefeller concept of the Energy Independence Authority with a bill passed in 1980 as the Energy Security Act, which established the Synthetic Fuels Corporation, backed with $88 billion. The corporation was to act as a private entity to finance projects for synthetic fuels to substitute for oil. The sources would be coal, shale, and tar sands. The Synfuels Corporation was to use price guarantees, purchase agreements, loan guarantees, direct loans, and joint ventures. Carter administration experts acknowledged that the initial costs would be in the range of $50 per barrel, nearly twice as expensive as the OPEC oil, but explained that once the corporation built a few plants to demonstrate that the technology was possible, OPEC would lower its price. The scheme upset both economists and environmentalists, however. Economists objected to distorting the efficiency of the free market with subsidies and government ownership. Environmentalists objected that mining millions of acres of coal, shale, and tar would destroy the land. After Reagan was inaugurated, he decided that the plan did not fit his free market policies and let it wither away without undertaking any new projects. Eventually its few early projects were abandoned also.

A mystifying feature of the last part of Carter's term was his decontrol of the price of crude oil. He spent the first two years of his term promoting the national energy plan, of which a key feature was regulation of the crude oil price. Originally imposed by Nixon and later put into law in the Energy Policy and Conservation Act of 1975, Congress reiterated it in the National Energy Act of 1978. The law did, however, have an escape clause (presumably for a military emergency) that the president could raise the price. Less than a year later, Carter announced that he would increase the amount gradually, so that within two years it would be decontrolled. Although this reversal of policy seemed perverse, it did have the desired effect of restoring efficiency. The higher prices caused industry and motorists to conserve fuel and brought more crude onto the market as marginal wells became profitable.

The Strategic Petroleum Reserve initiated under Ford might have seemed a natural program for Carter, with his practical engineering bent and his adherence to the New Deal style. Having half a billion barrels in the ground would discourage OPEC from another boycott. Nevertheless, the Department of Energy put very little oil into the salt domes. By the end of 1978 it contained only 67 million barrels, the equivalent of a mere eight days of imports. Part of the difficulty was technical, involving finding the right pumps and the right domes (the first one leaked). Some of the difficulty was price; the department did not want to pay the very highest prices from its limited budget. A few months later, the department stopped filling the reserve because Saudi Arabia was willing to increase its supply, but not if it were put into storage.

Carter's attempts to coordinate American policy with European allies and Japan were unsuccessful. In 1979 the price doubled to $38. That year proved to be a turning point because imports into the United States began to decline. Unfortunately the chief reason was the economic recession. High oil prices and high inflation (caused largely by oil prices) slowed demand considerably. Conservation measures began to take effect. Consumers were buying more efficient automobiles, and homeowners were insulating their houses. Industry was shifting from oil to coal.

In retrospect, many of Carter's energy policies appear inconsistent and even illogical. If deregulating the price of crude was desirable, why not put it in the national energy plan and the draft bill of the Energy Policy Act? Why incite the wrath of the environmentalists with the Synfuels Corporation that would destroy so much land? Why alienate economists and bankers with the Synfuels Corporation that would distort the free market? One reason, of course, is the New Deal tradition of the Democratic Party that believes that government intervention is desirable and that giant projects are beneficial. Carter did need the support of key Democrats in Congress, like Henry Jackson, who adhered to this view. Another reason is that due to his training and professional experience as an engineer, he gravitated toward technological solutions. Because of his naval duty on nuclear submarines, he was favorably inclined toward nuclear energy.

REAGAN AND THE RHETORIC OF THE LAISSEZ FAIRE MARKET

During his campaign for president in 1980, Ronald Reagan attacked Carter for his energy policy of conservation and government control. He told the voters that they did not need to sacrifice and that the problem was government regulation, and he promised to abolish the Department of Energy. Reagan, who professed a stronger ideology than any presidential candidate since Franklin Roosevelt, maintained that government intervention harmed the economy. His theories went counter to 40 years of Keynesian policy.

From Roosevelt's inauguration to the Nixon administration the interventionist theories of John Maynard Keynes dominated policy. Keynes argued that government needed to intervene by stimulating the economy when it was weak and moderating it when it was too strong. By themselves, supply and demand do not necessarily lead to full employment and stable prices. Often the result is unemployment or inflation. During the Great Depression, Roosevelt stimulated the economy by massive spending through the Works Progress Administration, Tennessee Valley Authority, and the Bonneville Power Administration. During World War II stimulus due to military production was far greater, and inflation threatened, so Roosevelt controlled wages directly. When John F. Kennedy campaigned for the presidency, the economy was weak after recessions in 1958 and 1960, so he promised to "get the country moving again." Once elected, he staffed his Council of Economic Advisers with Keynesian economists. Kennedy moved to stimulate the economy with a large tax cut, announcing that he did not believe the "myth [that] Federal deficits create inflation and budget surpluses prevent it."[34] The so-called Kennedy Tax Cut proved successful, leading to eight years of strong growth and high employment.

In general, the Democrats favored Keynes, and the Republicans were less enthusiastic. Keynes believed that the laissez faire market was inherently unstable and needed monitoring and correction, when the classical view was that it would naturally correct itself. Nixon was ambivalent, and fate was against him. High military spending for the war in Vietnam combined with increased social spending for the war against poverty had overstrained the economy, leading to inflation by 1970. His 1971 wage and price freeze departed sharply from the usual Republican policy. When questioned at a press conference, Nixon announced, "I am now a Keynesian."[35] The results were disastrous; the controls were unworkable, and so Nixon was left with continued inflation and a soiled reputation. As a practical man, the president quickly backtracked except for oil and construction.

Jimmy Carter shared much of the traditional Democratic ideology, incorporating both the New Deal technique of federal projects and the Keynesian intervention to control inflation and reduce unemployment. His national energy plan intruded deeply into private businesses. Yet at the same time, Carter was striving to deregulate business, first for airlines and next for trucking. Then he sought to deregulate railroads, securities, and banking.

Thus he was in the ironic position of both favoring and opposing the free market. The deregulation of airlines, trucking, and so forth led to efficiency, while the regulations and subsidies of the national energy plan, the Department of Energy, and the Synfuels Corporation led to inefficiency.

Ronald Reagan staunchly opposed the Carter administration regulations and advocated moving toward a laissez faire market as a means to increase efficiency. In this he was following the ideas of Milton Friedman, a University of Chicago economist. Like Adam Smith, Friedman believed in the classical view that the laissez faire market was the most efficient and would lead naturally to full employment and stable prices. He pointed to the golden age from 1815 to 1914, when the United States, England, and most of Europe had a free market and enjoyed great prosperity. World War I brought an end to laissez faire, and the Great Depression destroyed it. To Friedman, the combined inflation and high unemployment of the 1970s epitomized the worst of Keynesian interference. The remedy was to free business from the heavy hand of government and let the market solve the problem. Friedman's method was to concentrate solely on the money supply, which the Federal Reserve Board should allow to grow only in relation to productivity, optimally 3–5 percent per year.

Although Friedman held a minority position among economists, that position was widely known and had a following under the name of the Chicago School. He published two popular books[36] and wrote a column in *Newsweek* magazine for 18 years. He won the Nobel Prize in 1976, and in 1980 he starred in a 10-part television series titled *Free to Choose*. In the 1964 election he had advised Barry Goldwater, and he later advised Nixon until he imposed the wage and price freeze. Friedman first met Governor Ronald Reagan in 1970. The economist described the governor as a "very serious thoughtful person interested in principles" and that he "researched them through and by reading."[37]

Friedman and Reagan found further compatibility in the economist's assertion that economic freedom was the key to political freedom. He wrote that "economic freedom is an indispensable means toward the achievement of political freedom." Referring to government, he went on to warn that "by concentrating power in political hands, it is also a threat to freedom."[38] Friedman believed that the role of government should be limited to only a few functions, the chief being military defense. Other proper functions are enforcing contracts, preventing fraud, and stopping crime. Possibly the government needs to step in for two particular cases of market failure. The first is where a monopoly exists. The second is in the case of external costs, or as he phrases it, "neighborhood effects." Water pollution is a good example.[39]

Even then government should not be more intrusive than necessary and should weigh the disadvantages of its action in terms of cost and loss of freedom. But in general government's role should be only as an umpire calling the balls and strikes. Government decentralization is another tenet of Friedman's policy. As much as possible, functions should be at the lowest level. This makes it possible for a discontented citizen to vote with his feet, by moving to another city, county, or state. Friedman vehemently opposed central economic planning.

Reagan and his advisors eagerly embraced Friedman's brand of laissez faire. Taken as a whole, Reagan's policies acquired the popular title of Reaganomics. The president's first annual economic report read as if Friedman had been the author. It stated that "political freedom and economic freedom are closely related....All nations which have broad-based representative government and civil liberties have most of their economic activity organized by the market."[40] Regarding decentralization, the report said that "a reduced role for the Federal Government means an enhanced role for State and local governments....One constraint on the power of any government to impose costs on its citizens is the ability of those citizens to move elsewhere."[41] The actual authors of the report, of course, were neither President Reagan nor Friedman, but the Council of Economic Advisors.

With this clear ideology in favor of laissez faire economics, the Reagan administration seemed poised to be a natural experiment to test its supposed benefits over government planning and regulation. The new president moved this way eight days after his inauguration when he signed an executive order completely decontrolling the price of crude oil at the wellhead. He bragged of this in his first economic report, when he wrote:

My commitment to regulatory reform was made clear in one of my very first acts in office, when I accelerated the decontrol of crude oil prices and eliminated the cumbersome crude oil entitlements system. Only skeptics of the free market system are surprised by the results. For the first time in 10 years, crude oil production in the continental United States has begun to rise.[42]

Although the Reagan administration employed a rhetoric of the free market, this was only half the story. Many of the men who had supported Reagan and given him money were businessmen primarily devoted to their own self interest. Their praise for the market was combined with a desire to enhance their profits. Joseph Coors, for example, was a presidential intimate who considered OSHA inspectors a nuisance and EPA regulations controlling his brewery's wastewater an impediment. During the election many in the trucking industry had supported Reagan, and he received the endorsement of the Teamsters Union. Consequently Reagan stopped the deregulation of trucking, which had been progressing well under Carter. Although it would have led to greater efficiency, the Teamsters and many owners wanted to maintain their advantages under the old system. Their appeal to the new president was successful, and he pulled back. Vested interests won out over economic freedom and efficiency.

In terms of energy policy, Reagan did more than just decontrol crude oil and kill the Synfuels Corporation. He appointed members of the Federal Energy Regulatory Commission who eventually used their existing legal authority to deregulate the price of natural gas. His secretary of the interior, James Watt, strongly favored opportunities for private companies. At the department's Office of Surface Mining, without changing the law but only the implementation of it, the Watt team loosened environmental standards for mining coal, thus decreasing the operators' expenses. The Department of the Interior tried to increase greatly the amount of federal land leased for oil and gas production and coal mining. Although environmental groups managed to reduce the acreage, the department did lease much more than it would have under the Carter guidelines. The amount of OPEC oil imported dropped, but this was due to economic recessions in 1981 and 1982. Imports were only 4.3 million barrels per day compared to 8.6 million barrels per day in 1977. The Reagan administration's rejection of rationing and mandatory allocation during an emergency brought objections from the International Energy Agency, which pointed to the American treaty obligation in case of an emergency. The administration reaction was, "So what?" and it simply ignored the agency.

Energy policy was of minor concern to the Reagan administration, in contrast to the Carter administration, where it was primary. The logic was that a free market would stimulate the supply. Dampening demand was not mentioned particularly. This was certainly the opposite of planning. In fact, the policy worked fairly well. For five years, the price of oil on the world market stabilized, and then declined slightly in real terms adjusted for inflation. Then suddenly in 1986 it dropped from $27 to $13 a barrel in the course of a few months. The causes of this good fortune were multiple. Higher prices had made drilling for crude more profitable. Higher gasoline prices had dampened demand, and automobiles were more efficient in compliance with the 1978 law, a situation that took a number of years to take effect. OPEC could not maintain its monopoly due to new supplies from the North Sea and from Mexico. Nigeria was cheating on its OPEC partners because it needed cash. Thus the administration ideology of a free market was vindicated with respect to energy. In spite of promising in his campaign to abolish the Department of Energy, Reagan never did so. The Reagan administration had stepped up filling the Strategic Petroleum Reserve, and by the beginning of 1986 it had reached its original goal of half a billion barrels. Under pressure from Congress it continued to pump oil in, raising the goal to 750 million barrels.

The broader aspects of Reaganomics were more mixed in their results. To begin, the country suffered from a recession until 1982. Growth was flat for the first two years, and for all eight years of the Reagan administration, the rate was 3.4 percent in comparison to 5.2 percent during the expansion of the 1960s, often considered the success story for Keynesian planning. In comparison, the previous 11 years, with three recessions, had a growth rate of 2.4 percent. Two of those recessions can be blamed, at least in part, on the energy crisis. From 1950 to 1960, the fabulous fifties, the economy suffered from three recessions and had a growth rate of 4.2 percent. The most positive benefit of Reaganomics was to end the inflation that had been as high as 13 percent, and the most negative feature was the giant deficit in the federal budget, often more than $200 billion a year.

After the end of his administration, Reagan and his supporters could take pleasure in the collapse of the planned economies of the Soviet Union and Eastern Europe. Back in the 1930s, the Soviets had enjoyed better economic growth than the capitalist countries, which were suffering from the global Great Depression. New Deal planners had looked to the USSR as a model and had adopted features found in its five-year plans. (Indeed they had also looked favorably upon the Italian Fascist planning, but after World War II, never mentioned it.) By the 1980s the Soviet Union had accomplished building its big farms and heavy industries and needed to move on to consumer products, but its cumbersome system could not deal with these. Looking to the third world of Latin America, Africa, and Asia, Reagan found further proof of the benefits of the free market. These countries, many of whom had adopted central planning on the Soviet model, were stagnating. Ironically, some had adopted their plans with the encouragement of the U.S. foreign aid program. Milton Friedman considered India an especially good bad example.

In contrast to India, the Four Tigers of South Korea, Taiwan, Singapore, and Hong Kong grew at stupendous rates by following the opposite strategy beginning in the 1960s. They were too small to plan for balanced growth melding agriculture, mining, and heavy industry. Lacking these sectors, their key was exports. There was nothing to plan, only a lot of businessmen scrambling to sell the products of their factories to the United States and Europe. About the only contribution of their governments was to keep their hands off. Two other countries, Thailand and Malaysia, grew nearly as much. Across the oceans, American and European policy was favorable in that these industrial countries were in the process of reducing import tariffs.

These tigers were small, but starting a bit earlier, Japan had had even more startling economic growth, especially between 1955 and 1973 when industrial production increased 10 times. After a recession induced by the energy crisis, growth picked up again. Many pointed to the central planning of the Ministry of International Trade and Industry. More recently the Japanese miracle has faded. After the energy crisis, growth dropped from 10 percent annually to 4 percent, still quite good compared to the United States and Europe. The 1991 recession marked negative growth, and since then the rate has been about 1 percent, the lowest of the industrial world. Critics now point to the heavy hand of MITI.

Ronald Reagan took joy from the demise of economic planning in Great Britain. A year before his own election, Margaret Thatcher had become prime minister at the head of a Conservative Party government. Thatcher rejected most of the "socialistic" policies in place since the Labour Party victory in 1945. She proceeded to denationalize heavy industries like steel, coal, railways, and oil. She reduced the generous welfare benefits for housing, unemployment, and dependent children. Her belief was that these policies held back the British economy. As a result, the economy grew. Reagan and his advisors considered her policies models to emulate, and Thatcher was a frequent visitor to the White House.

THE 1991 GULF WAR

The last five years of the Reagan administration proved to be calmer than the first three, as the most zealous officials left, court decisions stymied some anti-environmental proposals, and opinion polls showed the public lacked enthusiasm for more Reaganomics. James Watt, the impassioned secretary of the interior, had been forced to resign after only a year and a half. In the 1988 election, when Vice President George H. W. Bush succeeded his boss, energy played little role in the campaign. Oil was $17 a barrel, about a third of the 1979 price (adjusted for inflation). Bush himself was a Texas oil man from Midland and Houston. After accumulating a fortune, he had devoted himself to government service, winning election to Congress, chairing the Republican National Committee, serving as ambassador to the United Nations and to the People's Republic of China, and directing the Central Intelligence Agency. Born and educated in New

England, he had enlisted during World War II and been a pilot in the South Pacific, earning a Distinguished Flying Cross and three Air Medals. And of course during his eight years as vice president, he had traveled widely, so he was personally acquainted with many foreign leaders. This cosmopolitan background gave him a greater experience in foreign policy than any modern president except Eisenhower. His preparation seemed ideal for the major challenge of his term in office.

On August 2, 1990, the tiny Persian Gulf country of Kuwait suffered a massive invasion from its longstanding enemy to its north, Iraq, under the leadership of its dictator, Saddam Hussein. Thirty thousand troops in three armored divisions rolled south and in five hours conquered the independent sheikdom the size of New Jersey. Its oil reserves amounted to 94 billion barrels, 10 percent of the world total. With a population of only two million people, its per capita income was among the highest in the world. Although officially a constitutional monarchy, its parliament was rudimentary, and democratic traditions were few. Most of the oil wealth went to the royal family and a few rich merchants, but there was enough to generously support the ordinary citizens, provided that they really were citizens. Over half of those living in Kuwait were aliens from other Arab countries, India, the Philippines, and elsewhere who did not share the wealth.

For propaganda purposes, Saddam Hussein's grievances were that the Kuwaitis were pumping oil from Iraqi reserves along the border, that they were flooding the market and thereby lowering the price, that the aristocracy violated Arab customs of equality, and that historically the area had been ruled as a province of Iraq until 1922 when the British intervened to make it an independent protectorate. A less acceptable reason was that Saddam Hussein desperately needed $30 billion to restore his country after 10 years of war with Iran. Indeed its long combat with Iran was a reason that the United States had looked favorably on Iraq. An enemy of Ayatollah Khomeini was a friend. Other reasons for American approval were that the country was fairly open and modern, that it tolerated Christianity, and that its disenchanted minorities like the Kurds threatened to begin a civil war that would destabilize peace in the region.

But within 24 hours President Bush decided that the Iraqi conquest could not be tolerated. The invasion violated international law. At first the president denied that access to Kuwaiti oil was an objective, but after a few days admitted that it was. He said that less availability and higher prices would harm the American economy. By temperament, George Bush was both conventional and impulsive. As a young man during World War II, he accepted the lesson that acquiescing to dictators led to war. The British and

French had given in to Hitler in his occupation of the Rhineland and in its occupation of the Czech Sudetenland after the Munich conference. The weak League of Nations had not stood up to the Italian dictator Mussolini when he invaded Ethiopia.[43] Virtually all presidents and diplomats during the Cold War shared this lesson of Munich and Ethiopia.

By coincidence, on August 2 Bush was attending an international conference on post–Cold War relations at Aspen, Colorado, where Margaret Thatcher, the prime minister of Great Britain, was to receive an award. After 11 years in power, she had won acclaim and reelection for revitalizing the British economy, reining in the welfare system, and winning the war against Argentina over the Falkland Islands. Like Bush, Thatcher fervently believed that appeasing Hitler at Munich had led to disaster. In discussing the invasion, she urged Bush to stand up to Saddam. Thatcher told him that Britain would support the United States militarily and assured him that France would to the same. In other words, the trans-Atlantic core of the alliance was already in place.

Reflecting back on his presidency, George Bush considered organizing the coalition of 40 nations that contributed 200,000 troops to be the acme on his career. Much of the credit goes to him personally because he continuously telephoned to persuade prime ministers, presidents, and kings around the world, most of whom he knew individually. Margaret Thatcher did not need persuading; indeed she did much to persuade him. Germany, Italy, and France joined quickly. The European Community immediately froze Iraqi money in international banks. Japan, which imported 12 percent of its oil from Iraq, agreed to participate. Turkey agreed to support the emerging coalition and stopped the oil flowing in the pipeline from Iraq. The collapse of the Soviet empire in Eastern Europe less than a year earlier produced a diplomatic revolution. Following a visit by Secretary of State James Baker, the USSR agreed to cooperate on economic sanctions. The People's Republic of China did the same. On August 6 the UN Security Council voted to impose economic sanctions. Thus the industrial world and others took a formal stand in an international organization.

Organizing the Arabs looked difficult at first. Due to their hatred of Israel, they had never cooperated collectively with the United States. Individually, however, many of them were closely tied to America. Moreover, most of them feared Saddam. The conservative regime of Saudi Arabia had the most to lose. It feared Saddam would continue his attack across the Kuwaiti border into its territory, seizing the rich oil fields in the eastern province of Al-Hassa. The small Saudi army would have been no match against the Iraqis. The tiny Gulf kingdoms of Bahrain, Qatar, the United

Arab Emirates, and Oman trembled before Saddam. Egypt signed with the Americans. Its moderate government had no love of Saddam and received $2 billion a year in aid from the United States. Within a few months the United States arranged to forgive hundreds of million dollars of debt. Syria, under the brutal dictator Hafiz al-Asad, was no friend of the United States, but also bitterly hated its neighbor to the east, and hence cooperated.

Mobilizing the troops and firepower in northern Saudi Arabia against the Iraqis in Kuwait proved difficult. Eventually the United States deployed 540,000 troops, and allies deployed 250,000. The location was 8,000 miles from home, and remote from any overseas base. The desert was hot, water was scarce, and the sand infiltrated the artillery, tanks, and electronic equipment. Operation Desert Shield, the code name for the defense of Saudi Arabia, had 250,000 troops in the field by October, but this was too few for a successful attack to expel the Iraqi army from Kuwait. The chairman of the Joint Chiefs of Staff, General Colin Powell, reflected the post-Vietnam doctrine of the military about the importance of having enough troops and firepower to assure victory, in contrast to the piecemeal buildup of troops in Vietnam, where the United States never had enough. When in October the president asked for a formal assessment of liberating Kuwait, General Powell and the commander in the field, General Norman Schwarz-kopf, candidly presented a plan that predicted almost no chance of victory, causing Bush to order the number of troops doubled.

On January 17, the air force began dropping a total of 142,000 tons of high explosives, until Iraq was a wasteland. Then on February 24, ground troops invaded, making a wide flanking attack to the west. After a hundred hours, Bush called off the attack since all organized resistance had ended. Only 148 Americans were killed, compared to 100,000 Iraqis. Saddam's army lost 3,700 tanks and the Americans lost zero. Seven hundred fifty oil wells caught fire, consuming six million barrels a day. Incredibly Saddam Hussein clung to power in Baghdad. In spite of his military losses and the suffering of the civilian population, the dictator did not flee or suffer a coup d'état as had been expected. Bush's reasons for ending the attack on February 27 were to not exceed the agreement of the coalition, which was only to liberate Kuwait, to prevent a slaughter of troops fleeing to Baghdad, to avoid having to occupy Iraq, and because he expected Saddam Hussein to be overthrown.

After the victory in the Gulf War, the price of oil dropped and the attention paid to energy also dropped. The year prior to the invasion, oil sold for $18 a barrel, and the year after the invasion it sold for $18. Although the spot price shot up to $40 just prior to the air attack in January, the average price for the whole year was $21, only $3 higher than before and after. The big winner appeared to be George Bush. His public opinion approval rating on the Gallop Poll shot up to 83 percent, the highest of any president up to that time. He seemed invincible for reelection in 1992. The cloud on the horizon that eventually did much to cause his defeat was the economic recession that began in 1990 and continued until 1992. Some observers trace the cause to higher energy prices due to the invasion and war.

THE GEORGE W. BUSH ADMINISTRATION

In 1992 Bill Clinton campaigned hard on the issue of the economic recession, and Bush (Senior) lost in spite of riding so high in the polls the year before. Once elected Clinton devoted almost no attention to energy policy. The recession ended, and an expansion began that continued to be the longest in history. Energy issues slipped from sight. The price of oil remained low, and since the prices of other fuels follow the price of oil, little happened. Ironically a Democrat was adhering to the Reagan policy. He did not engage in the planning associated with the New Deal and the Carter administration.

The very first energy issue to confront the new administration of George W. Bush was the electric blackout in California. At nearly the same time as he was taking the oath of office, the Pacific Gas and Electric and Southern California Edison companies were cutting power to millions of businesses and homes. Suddenly Silicon Valley, the most industrially advanced region in the most industrialized country, faced shortages comparable to the third world. At first glance the problem seemed to be entirely caused by government. In the 1992 Energy Policy Act, Congress had directed the state public utility commissions to deregulate the industry.

Although all states moved in this direction, California was in the forefront. Its legislature passed its deregulation law in 1996, one of the earliest ones. At the time, the state had lots of power (30 percent more capacity than it needed). Its large population of 34 million and prosperous economy seemed to offer a healthy market. On the other hand, California depended heavily on hydropower, which varies with rainfall, and imported much electricity from Oregon, Washington, and Arizona. Due to its remote location, the state's connections to the national grid were weak. Its growing population and economy meant demand had to increase quickly. Severe air pollution precluded coal and oil plants. An unexpected jump in the cost of natural gas drove up generation costs. Furthermore, the state was the victim of massive fraud, although unrecognized at the time.

At the time, most of the failure appeared to be due to governmental decisions. The state Public Utility Commission enthusiastically championed deregulation, but in a heavy-handed way that was not truly a free market. Indeed it was a hybrid restricting the market with disastrous central management. As in all other states, the electric industry was integrated vertically for generation, transmission, and distribution. The law left distribution the same, but changed generation and transmission. Previously the utilities owned the plants, and the PUC set the prices. Afterward, the utility companies sold most of their plants and the prices were set by the Power Exchange, a nonprofit organization. Previously the utilities owned and operated the transmission lines and power grid. Afterward, the Independent System Operator, a nonprofit organization, managed the system.

The PUC only deregulated half of the system. It demanded that the utilities divest themselves of their power plants and demanded that they buy electricity on the open market. Indeed it even forbade them from entering into long-term contracts with suppliers. They were obliged to buy on the spot market. The rationale was that the spot market would be cheaper, an assumption that proved to be disastrous. For the other half of the equation, the price charged to consumers, the PUC required the utilities to not increase prices until 2002. The commission's financial squeeze combined with many years of official policy to discourage construction of generating plants. Much of the opposition was for safety and environmental protection. The public had demanded that the nuclear plant at Diablo Cliffs be shut down when scientists realized it was built on an earthquake fault. Repairs sent the cost from $620 million to $5.8 billion. The nuclear plant at Humboldt Bay closed in 1976 after only 13 years of operation when an earthquake fault was discovered beneath it. The consequence of blunders and environmental caution was that virtually no new generating capacity had been built in the Golden State for the prior 10 years.

With the shortage of electricity in California, the Democratic governor, Gray Davis, appealed to the newly inaugurated president, who showed little sympathy. After admonishing the state for trying to repeal the laws of economics, the new president merely directed the Department of Energy to require suppliers in the West to sell to California for two weeks until the state could turn around its system. This was very little time. Treasury Secretary Paul O'Neill lamented that the state was trying to "defeat economics" by freeing wholesale prices while freezing consumers prices, labeling it "lunacy." In spite of the crisis, Governor Davis still wanted to continue the price cap for consumers. The cost to the state to bail out the system was in the range of $20 billion. The cost to consumers jumped 40 percent.

Pacific Gas and Electric filed for bankruptcy. Much blame attached to the governor, culminating two years later in a recall election. The citizens of California considered that Davis had blundered so badly on the electricity crisis, combined with fiscal problems of the state, that they voted him out of office and replaced him with Arnold Schwarzenegger, a former body-builder, actor, and businessman.

The fraud that contributed to the crisis, unknown at the time, was manipulation of the wholesale price by the Enron Corporation. This company had greatly expanded its scope under the 1992 law as a broker buying from exempt wholesale generators and selling to distributors. Its deceitful technique was to make phony transactions between different parts of the corporation that were supposedly independent, which yielded an inflated price, which it then charged to California distributors. Once the scheme was discovered, Enron went bankrupt. Top executives pled guilty to fraud. Other companies the state blamed for the crisis were Duke Energy, Mirant, and Williams.

Soon after his inauguration, and partially prompted by the California crisis, Bush directed Vice President Dick Cheney to develop a national energy plan. The idea came as a surprise to many, since it smacked of central planning not seen since the Carter administration. Cheney worked rapidly, seeking advice from industry leaders, a move that proved controversial when he refused to make their identities public. Critics charged that this was an attempt to favor industry over the interests of the American consumer. Bush had received generous donations from petroleum and electric company executives during the presidential campaign. Officials from Enron gave $1.8 million. Exxon executives donated $1.2 million, and BP Amoco executives donated $800,000. Other major contributors were from El Paso Energy, Chevron, and the Southern electric company.[44] Since Cheney had recently been head of the Halliburton Company, an oil company, and Bush had been in the oil business early in his career, the energy plan received a lot of criticism. Environmentalists were quick to call Cheney's proposal a payback.

The controversial elements of the plan were to greatly increase burning coal, increase nuclear generation, and to open Alaska to more exploration and production. It ignored opportunities to save energy by conservation. During the election campaign Bush had proposed drilling in the Arctic National Wildlife Refuge, suggesting that the area might hold as much oil as the Prudhoe Bay area. The energy plan also proposed expanding leasing in the National Petroleum Reserve to the west of Prudhoe Bay. It advocated building a pipeline across Canada to bring natural gas from Alaska. In the lower

48, it proposed drilling for oil and gas in coastal regions and on the Outer Continental Shelf. Tax breaks for drilling were supposed to encourage more production. Regarding electricity, Cheney's group recommended improving the electric grid. To promote burning coal, the energy plan recommended $2 billion in research, more leasing on federal land, and relaxed pollution standards. Benign terminology for this included directing government agencies to "provide regulatory certainty," "expedite permits," and "reexamine the current federal legal and policy regime."[45] Besides this veiled vocabulary, the report contained many measures that appeared pro-environmental like weatherization, conservation, and renewable energy. In terms of rhetoric, the report echoed many proposals of the 1977 national energy plan.

The tragedy of September 11, 2001, reordered the Bush presidency. In the face of the Al Qaeda attack on the World Trade Center and the Pentagon, the public rallied around him, pushing his popular opinion approval rating to 92 percent. A few months earlier, it had been 52 percent. It soon became apparent that Al Qaeda, under the leadership of Osama bin Laden, was operating out of Afghanistan, under the aegis of the extreme Muslim Taliban government. Bush declared a war on terrorism. His team began plans to attack, and on October 7 the air force and the navy began heavy bombing. Two weeks later army rangers were on the ground in combat. More American forces arrived and soon linked up with Afghans opposed to the Taliban. On November 12 the capital city of Kabul fell, and on December 21 an interim government was installed. Outside the cities, guerilla attacks diminished, although Bin Laden himself escaped. The quick military victory heartened the United States.

The success of the war in Afghanistan next pointed the Bush administration toward military action against Iraq and its dictator, Saddam Hussein. The situation there had been a sore point with certain members of the president's team dating back to the end of the Gulf War. A number of his staff had served in his father's administration. Moreover the new president himself felt antipathy toward Saddam due to the Gulf War and a later Iraqi attempt to assassinate his father. In the last few days of the Gulf War, President Bush (Senior) had called off the attacks against the retreating troops. Hindsight proved that decision to be a mistake. Saddam did not fall, and his regime did not relent in its hostility toward moderate Arab neighbors and toward Israel. Furthermore Saddam brutally attacked and killed Kurdish and Shiite people within his own country who opposed his dictatorship. Some observers speculated that he was continuing to try to build an atomic weapon even though the Israeli air force had bombed the experimental Osirak reactor in 1981.

A leading advocate of attacking Iraq was Paul Wolfowitz, the deputy secretary of defense. He had floated the idea at a meeting at the presidential retreat at Camp David the weekend after September 11. Ten years earlier, Wolfowitz had served in the Pentagon of the president's father in the capacity of undersecretary of defense for policy, where he had played a key role in the strategy for the Gulf War. He believed it was a mistake not to have removed Saddam from power then. Trained as a political scientist, Wolfowitz had had a long career in government, serving in the Arms Control and Disarmament Agency and the Pentagon during the 1970s and heading the State Department Policy Planning Staff during the Reagan administration. While out of government during the Clinton years, he had continued to decry the Iraqi regime. His boss, Secretary of Defense Donald Rumsfeld, also advocated an attack to remove Saddam and was convinced by the quick success in Afghanistan that a military invasion of Iraq would be comparatively easy.

The large oil reserves of Iraq were a strong argument to liberate the country. Its official tally of proven reserves is 112 billion barrels, second only to Saudi Arabia. As a consequence of the 1991 war, Iraq was restricted in the amount it could export. If this could be returned to full production and come into friendly hands, the advantage to the economies of the United States and its allies would be great. In candidly explaining the situation sometime later, Wolfowitz said, "The country swims on a sea of oil."[46]

President Bush did not publicly use this argument in trying to persuade Congress and the American people to back him. Instead he pointed to connections he saw between Iraq and Al Qaeda, and the risks of weapons of mass destruction. The Al Qaeda connection was supposed to be based on a visit by an operative to Iraq and on meetings in the Czech Republic. This information was elusive and eventually discredited. "Weapons of mass destruction" was a term introduced to encompass poison gas, disease pathogens, and nuclear weapons. Beyond doubt Saddam had possessed poison gas, which he had used against the Iranians during his war against them in 1980–88 and had used on his own citizens who opposed his regime. His capability for using them against the United States, however, was nil. He lacked the missiles or aircraft to deliver them across the Atlantic Ocean. Even using poison gas against Israel seemed unlikely, since his missiles were so bad. Biological attack appeared to have some basis in fact because of the anthrax letters sent to members of Congress immediately after September 11. About eight postal workers, reporters, and others contracted the disease and died. But virtually no evidence existed that the source was Al Qaeda, and the likelihood appeared to be that it was fiendish sabotage by a domestic terrorist. Disease pathogens require advanced medical skills and scientific laboratories, are dangerous to those who prepare them, and are hard to maintain and ship, thus making them an unlikely mass danger. At most, they would infect only a few dozen victims.

Nuclear weapons are a different story, however. Even a small bomb could kill tens of thousands. The one dropped on Hiroshima killed 70,000 immediately and another 70,000 due to radiation over the following weeks. Vice President Dick Cheney warned of a Saddam "armed with an arsenal of these weapons of terror" who could "directly threaten America's friends throughout the region and subject the United States or any other nation to nuclear blackmail."[47] Yet Iraq had no civilian reactors that could be converted to manufacture bombs secretly. The administration speculated that the Iraqis had an independent program. Two items (later discredited) were its purchase of high-strength aluminum tubes that could be used to enrich the uranium and negotiations to purchase uranium from Niger in Africa.

Bush continued to push for war. At the United Nations, the United States got little support except from Britain, so it dropped its attempts to get a UN resolution. France and Germany were totally opposed to helping an invasion. At home, Congress voted to support a war. Republicans backed it overwhelmingly, while Democrats were ambivalent. A number of senators, such as Joe Lieberman and Hillary Clinton, voted in favor, worrying that the war fever was so hot that voters would punish politicians who did not favor an invasion. (Both harbored presidential ambitions.) Other Democrats voted against it because they did not believe the evidence of weapons of mass destruction, because U.S. and British troops were too few for an invasion and occupation, and because it violated the American tradition of seeking peaceful solutions. Beginning in January many citizens held demonstrations against the attack. On March 19 American and British warplanes began bombing. The next day troops positioned in Kuwait invaded and advanced rapidly. By April 9 troops entered Baghdad and in another week controlled nearly the entire country. As with Afghanistan, the military victory came quickly.

The fighting itself damaged the wells, pipelines, and refineries, but four times as much damage came after the war from sabotage and insurgent attacks. Furthermore 12 years of neglect had reduced production. Equipment was old, pipelines were rusty, and skilled personnel were scarce. Moreover, some geologists disputed the amount of oil reserves in Iraq. The number had been arbitrarily raised to 112 million barrels at a time when the country wanted to secure a foreign loan. Recent estimates vary greatly. The U.S. Geological Survey estimated that Iraq has only 45 billion barrels of reserves, while the Iraqi oil minister estimated 214 billion barrels.[48]

With the defeat of Saddam Hussein, the United States found itself in charge of the government for a year, thus confronting the question of how much it should rely on central planning. The Bush administration certainly did not intend to establish a socialist system, but Saddam's regime under the Ba'ath Party had been socialistic, at least in theory. In fact much was merely an excuse for dictatorship. Within the United States, some considered the liberation of Iraq to be an opportunity for American corporations to gain access to the oil industry. In the earliest days of the occupation, the Pentagon awarded contracts to oil companies like Halliburton (formerly headed by Dick Cheney) to rebuild the industry. Public outcries led to some contracts being revoked and given to Iraqis like the Southern Oil Company based in Basra.

Looking worldwide, geologist Craig Hatfield has argued that virtually no new reserves have been added to the global supply since the 1970s. By that time geologists had explored the entire globe from the Arctic to the Southern hemisphere. There is no space left on earth for a super-giant field such as Alaska or the Middle East. Geologists have looked everywhere. The few fields discovered recently, such as those in the former Soviet Union, have been of only moderate size. Indeed, fields such as the Caspian Sea have been known in a rough way since 1910. Hatfield acknowledged that in 1987 Saudi Arabia and Abu Dhabi officially raised their estimated reserves by 277 billion barrels, but he countered that these revisions were phony, undertaken to secure big loans from international banks. Hatfield went on to observe that studies of other minerals have shown that once depletion has brought reserves down to half their maximums, prices tend to rise dramatically, and he estimated that this may occur as early as 2011.[49] He further concluded that substitution of synthetic fuel is not practical. To do so would require building 40 plants at the cost of $20 billion each in the next decade. Even if the money were available, the engineering technology is not.

The demand side of the economic equation is ominous too. Demand from the third world is surging. The People's Republic of China, with a population of 1.3 billion, has extreme potential. Its growth rate has been 9 percent recently. It is already buying up oil fields in Kazakhstan in the former Soviet Union. India, with a population of one billion, has a growth rate of 7 percent. After a brief recession in 1997–99, the Asian Tigers of South Korea, Taiwan, Singapore, Thailand, and Indonesia are growing at a rate of nearly 5 percent. Within the United States, two energy trends point in opposite directions. On one hand, the amount of energy used per unit of output continues to decline. Efficiency is greater. On the other hand, the total demand increases with prosperity, causing more demand for automobiles, jet travel, and big houses. Moreover, urban sprawl and bigger houses increases demand for gasoline and heating. So in spite of greater energy efficiency, overall consumption increases.

CONCLUSION

Jimmy Carter, the most religious president in recent years, took on the mantle of a prophet in his speeches about energy, calling the crisis "the moral equivalent of war." His jeremiad envisioned it as a consequence of a "national malaise." Earlier, President Nixon also had compared it to war and invoked his authority derived from World War II mobilization legislation. From the secular side, Americans had heard many cries of doom over the years. The 1908 White House Conference on Natural Resources had pointed out coming shortages. The Paley Commission had recommended to President Truman that the United States should have "a comprehensive energy policy." M. King Hubbert warned in 1956 that oil reserves would peak in the late 1960s. And in 1972 the Club of Rome predicted an energy crisis. Every one of these was ignored. All of these predictions of an energy shortage depended on statistical facts, and the club was an innovator in analysis.

Over the decades there have been a number of plans to address energy shortages. Nixon had to deal with the oil embargo as an emergency, so it is hard to expect that he have a plan, at least at the beginning. He did act within weeks to create the Federal Energy Office. For his legal authority to allocate gasoline, he used legislation originally passed for wartime mobilization. A major addition was his plan for an energy independence authority, again following a war model. Carter had more time for comprehensive planning, which he called the national energy plan. Like the Nixon and Ford policies, it depended on command rather than the market to control the supply and demand. While it was comprehensive, it was not internally consistent. For example it discouraged bringing new supplies of crude oil on stream. The failures of the Nixon and Carter plans make it ironic that, 24 years later, President George W. Bush had Dick Cheney set forth an energy plan.

Taking the opposite tack to Carter (and to Nixon), Ronald Reagan rejected the war mobilization model in favor of a free market, first for crude oil and a few years later for natural gas. This was the antithesis of central planning and proved successful.

NOTES

1. Richard Nixon, "Special Message to the Congress on the Energy Crisis," January 23, 1974, *Public Papers of Richard Nixon 1974* (Office of the Federal Register, National Archives and Records Administration).

2. Donella H. Meadows, Dennis L. Meadows, Jorgen Randers, and William W. Behrens III, *The Limits to Growth* (New York: Universe Books, 1972), Table 4, Nonrenewable Natural Resources, p. 58.

3. Christopher Flavin and Nicholas Lenssen, "Designing a Sustainable Energy System," in *State of the World,* ed. Lester R. Brown et al. (New York: W. W. Norton, 1991), p. 20.

4. Arden Bucholz, *Moltke, Schlieffen, and Prussian War Planning* (New York: Berg, 1991), pp. 20–21, 28, and 47.

5. Ibid., p. 278.

6. Paul A. C. Koistinen, *Mobilizing for Modern War* (Lawrence: University Press of Kansas, 1997), p. 236.

7. Robert Engler, *The Politics of Oil* (New York: MacMillan, 1961), pp. 132–133; Gerald D. Nash, *United States Oil Policy 1990–1964* (Pittsburgh: University of Pittsburgh Press, 1968), pp. 23–48; and Harold F. Williamson, et al. *The American Petroleum Industry* (Evanston, IL: Northwestern, 1963), pp. 4–14.

8. Robert F. Himmelberg, *The Origins of the National Recovery Administration: Business, Government and the Trade Association Issue 1921–1933* (New York: Fordham, 1993), p. 45.

9. Charles F. Roos, *NRA Economic Planning* (New York: Da Capo Press, 1971, originally 1937), p. 14.

10. Quoted in Roos, *NRA Economic Planning,* p. 6.

11. Roos, *NRA Economic Planning,* p. 45.

12. Ibid., p. 63.

13. Ibid., pp. 355–359.

14. Ibid., pp. 471–472.

15. Interstate Oil and Gas Compact Commission (Oklahoma City: IOGCC, 1999).

16. Marion Clawson, *New Deal Planning: The National Resources Planning Board* (Baltimore: Resources for the Future, Johns Hopkins University Press, 1981), pp. 60–63.

17. Ibid., pp. 259, 178–179.

18. Otis L. Graham, *Toward a Planned Society: from Roosevelt to Nixon* (New York: Oxford, 1976), pp. 52, 59, 62.

19. Ralph J. Watkins and Wilbert G. Fritz, "Planning for Energy Resources," in *Planning for America,* ed. George B. Galloway (New York: Henry Holt and Co., 1941), pp. 139 and 142.

20. Herman Miles Somers, *Presidential Agency OWMR* (New York: Greenwood, 1969, originally 1950), pp. 6, 7, and 10.

21. Ibid., p. 33.

22. Ibid., pp. 78, 81.

23. Henry L. Trewhitt, *McNamara* (New York: Harper and Rowe, 1971), pp. 36–41; John A. Byrne, *The Whiz Kids* (New York: Doubleday, 1993).

24. National Petroleum Council, *U.S. Energy Outlook,* December 1972 (actually released in March 1973).

25. V. R. Cardozier, *Mobilization of the United States in World War II* (Jefferson, NC: McFarland, 1995), pp. 104–105.

26. Henry Kissinger, *Years of Upheaval* (Boston: Little Brown, 1982), pp. 856–857.

27. William E. Simon, *A Time for Truth* (New York: McGraw Hill, 1978), p. 51.

28. Ibid., p. 53.

29. Ibid., p. 55.

30. Richard Nixon, *Memoirs* (New York: Grosset and Dunlap, 1978), p. 786.

31. Robert J. Leiber, *The Oil Decade* (New York: Praeger, 1983), p. 24.

32. Michael Turner, *The Vice President as Policy Maker* (Westport, CT: Greenwood, 1982), p. 101.

33. Executive Office of the President, *The National Energy Plan* (1977).

34. Elton Rayack, *Not So Free to Choose* (New York: Praeger, 1987), p. 4.

35. Ibid., p. 6.

36. Milton Friedman, *Capitalism and Freedom* (Chicago: University of Chicago Press, 1962); and Milton Friedman with Rose Friedman, *Free to Choose* (New York: Harcourt Brace Jovanovich, 1980).

37. Rayack, *Not So Free,* p. 1.

38. Friedman, *Capitalism and Freedom,* pp. 8 and 2.

39. Ibid., p. 25.

40. Council of Economic Advisors in the Executive Office of the President, *Economic Report of the President,* February 1982, p. 27.

41. Ibid., pp. 5–6, 36.

42. Ibid., p. 7.

43. George Bush and Brent Scowcroft, *A World Transformed* (New York: Knopf, 1998), pp. 340 and 303.

44. Greg Palast, "Bush Energy Plan: Policy or Payback," BBC News, May 18, 2001.

45. White House, National Energy Policy Development Group, *National Energy Policy,* May 2001.

46. George Wright, "Wolfowitz: Iraq War Was about Oil," *The Guardian,* June 4, 2003.

47. John B. Judis and Spencer Ackerman, "The Selling of the Iraq War," *New Republic,* June 30, 2003.

48. Energy Information Administration, "Country Analysis Brief: Iraq," June 2006, www.eia.doe. gov/emeu/cabs/Iraq/Oil.html.

49. Craig Hatfield, "Oil Back on the Global Agenda," *Nature,* May 8, 1997, p. 121.

— 4 —
Overpopulation: The Absence of Planning

Writing in 1968, the biologist Paul Ehrlich warned that "hundreds of millions of people are going to starve to death." We must have "determined and successful efforts at population control.... We must have population control at home.... We must push other countries into programs...."[1] Ehrlich titled his book *The Population Bomb* to equate the problem to a ticking time bomb about to explode. When Ehrlich wrote, the world had three billion people, but by 2007 it had more than six billion.

Ehrlich described how, in prehistoric time, the world population was stable until the end of the New Stone Age about 8,000 years ago, when humans learned to cultivate crops, thereby greatly increasing their food supply. Archeologists estimated that the world population was 5 million, increasing to 50 million by 1000 B.C. Life was hard and the death rate was high, a condition similar to underdeveloped countries until about 1950. By the time of the Roman Empire, population reached 300 million, where it stayed for more than a thousand years through the Dark and Middle Ages. By 1450 improved agriculture and less warfare in Europe brought about an explosion to 500 million worldwide, and by the early years of the Industrial Revolution in 1800 it climbed to one billion. As the European and New World standard of living increased dramatically, so did the number of people. By 1927 it had doubled again to two billion. Improvements in medical science in Europe and North America cut the death rate greatly, and better health and more wealth encouraged people to have more babies.

After 1945 the industrial countries extended the benefits of modern medicine to the underdeveloped countries. DDT sprayed to control malaria-carrying mosquitoes was one of the most successful techniques. The program in Ceylon (now called Sri Lanka) demonstrated its effectiveness. Prior to spraying DDT in 1946, the overall death rate was 22 per 1000, but by 1954 it was less than half: 10 per 1000. Around the world, underdeveloped countries enjoyed medical victories over yellow fever, smallpox, and cholera that caused mortality to plunge.[2] As the death rate declined and the birth rate continued high, the population exploded.

The immediate problem was a shortage of food. For the first decade after World War II agricultural production in the underdeveloped countries managed to keep up with population. Then about 1958 "the stork passed the plow," colorful language to indicate that births exceeded agricultural capacity. Food prices rose, and farmers had to cultivate marginal land. The increasing food production could not keep up with the increasing number of mouths to feed.

To solve the problem of the population bomb, Ehrlich proposed a dozen practical remedies. He considered the United States the place to start, because with 6 percent of the world population, it consumed 80 percent of its resources, according to his figures. Until America had a population policy in place, it could not persuade other countries to do the same. The first step would be to change the income tax law that rewarded having children with deductions to penalties of $1,200 per child. Parents without children would get prizes. Next the government should establish the Department of Population and Environment to educate, persuade, and conduct research on birth control, including mass sterilization. Education would teach the benefits of recreational sex, so that "we should be able in a generation to have a population thoroughly enjoying its sexual activity.... relatively free of the horrors created today by divorce, illegal abortion, venereal disease, and the psychological pressures of a sexually repressive and repressed society."[3] Ehrlich considered the anticontraception position of the Roman Catholic Church to be a major impediment and recommended that "we must bring pressure to bear on the Pope in hope of getting a reversal of the Church's position."[4]

Once the United States had adopted sane policies at home, Ehrlich said it would need to find a solution for the rest of the world. He had abandoned hope for large regions of the globe being possible to save, including India. American wheat would only go to places with the agricultural potential to feed themselves. He foresaw a crisis period of about 10 years when the underdeveloped countries with potential would need to receive American food, sex education, and agricultural assistance. Television sets in rural villages would receive broadcasts on desired family size and contraception. Farm extension agents would introduce improved crops and farming techniques. Ehrlich believed that after a decade these countries would have turned the corner with smaller populations and more food. On a broader picture, scientists would need to determine the optimal population for the world. Ehrlich speculated it would be in the range of one or two billion.[5]

As its author intended, *The Population Bomb* created an explosion when published. It sold three million copies. Its most alarmist predictions did not occur quite as fast as Ehrlich predicted, however. Although mass starvation did not occur worldwide, it did occur regionally. By the end of the 1970s, millions starved as the Sahel region in West Africa suffered a drought caused by overpopulation. Millions more starved in Bangladesh. Twenty-two years later, Ehrlich wrote a sequel, titled *The Population Explosion,* which began by saying:

In 1968, *The Population Bomb* warned of impending disaster if the population explosion was not brought under control. Then the fuse was burning; now the population bomb has detonated. Since 1968, at least 200 million people (mostly children) have perished needlessly of hunger and hunger related diseases.[6]

Ehrlich lamented that humans are genetically programmed to reproduce, hence resist the simple scientific facts about its dangers. For instance, Costa Rica has an excellent voluntary family-planning program combining education and contraception, but has an average family size of 3½ children, and a growth rate that will double the population in 20 years. Ehrlich praised the Chinese one-child policy as a necessity. It quickly reduced the birth rate. India, on the other hand, has failed and is making the biggest contribution to overpopulation. Africa he described as "a demographic basket case." Kenya is the worst, having doubled its size and actually increased its rate to a completed family size of eight. Ehrlich wrote negatively about the Moslem world for its high birth rate, which he attributed to the low status of women.[7]

Timeline for Overpopulation

534	Bubonic plague kills 70,000 in Constantinople, then spreads to Italy and France.
1347	Black Death begins killing one-quarter to one-third of the population of Europe.

1492	Columbus discovers America. The great Atlantic migration begins.
1607	English found Jamestown, their first permanent colony in North America.
1632	John Graunt begins to analyze births and deaths in London.
1664	Jean Baptiste Colbert establishes the French East India and West India Companies, marking a high point of mercantilism.
1790	Congress passes the Naturalization Act requiring two years residence for citizenship.
1798	Thomas Robert Malthus published his *Essay on Population*.
1849	American Party founded, also called the Know Nothings, who opposed immigration.
1865	Immigration picks up again after the Civil War. 25 million come by 1914. Francis Galton coins the term eugenics.
1882	Chinese Exclusion Act.
1907	Gentleman's Agreement prevents immigration from Japan.
1912	Margaret Sanger works full time on birth control in New York City.
1914	Immigration ends abruptly with outbreak of World War I in Europe.
1921	Restrictions on immigration are made permanent in the Johnson Reed Act of 1924, based on national quotas.
1927	World Population Conference in Switzerland. Supreme Court permits involuntary sterilization.
1939	Planned Parenthood Federation organized.
1948	Congress passes the Displaced Persons Act admitting 400,000.
1954	World Population Conference in Rome.
1955	People's Republic of China announces birth control.
1956	Pres. Eisenhower admits 30,000 Hungarian refugees.
1958	Draper Committee urges foreign aid for population control in underdeveloped countries.
1960	Birth control pill invented and on the market.
1961	India begins birth control.
1965	Immigration Act ends national quotas.
1973	*Roe v. Wade* makes abortion legal in all states.
1974	World Population Conference in Bucharest.
1975	State of emergency in India emphasizes birth control by sterilization.
1976	Congress passes Hyde Amendment prohibiting government payment for abortions.
1978	China strictly enforces the one-child policy.
1984	World Population Conference in Mexico City. Reagan administration announces the global gag rule.
1986	Simpson Rodino Immigration Act addresses problem of illegal aliens. National Right to Life Committee founded.
1993	Pres. Clinton rescinds the global gag rule.
1994	World Population Conference in Cairo.
2001	Pres. George W. Bush reinstates the global gag rule.

FROM THE ANCIENT WORLD TO
AMERICAN COLONIZATION

Like other nations in the cradle of civilization, the ancient Hebrews believed that having many descendants was a sign of divine favor. In the Garden of Eden, God directed Adam and Eve, "Be fruitful and multiply, and fill the earth...." "After the great flood, God promised Noah that his descendants would people the whole earth. Ten generations later, when Abraham was distressed that he did not have a legitimate son by his wife Sarah, God promised that his descendants would be as numerous as the stars in the heavens.[8]

Looking back to prehistoric time, demographers point to the neolithic transition about 8,000 or 10,000 B.C. when humans shifted from hunting and gathering to farming. Agriculture became far more efficient in supplying food, yet the life expectancy decreased from 19 years to 17 years. Skeletal remains show that body size, height, and bone thickness decreased. The cause was a worse diet and more disease due to crowding. Hunter–gathers ate a balanced diet of roots, greens, berries, and game, whereas farmers ate chiefly grains. With the transition to farming, reproductive strategy changed as well, as women began to have more children. The world population grew from 5–6 million to 250 million in the year A.D. 1. Although the total number was great, the annual rate of increase was only 0.37 per 1,000, less than a tenth of the rate in many third–world countries today.[9]

Plato discussed population policy in the *Republic*, where he recommended an optimal size for the city-state of 5,040 citizens, with fertility restrained by conscious birth control. Methods were late marriages, prostitution, coitus interruptus, homosexuality, abortion, and infanticide. In addition, both Plato and Aristotle recommended destroying deformed babies. The decision was to be made by the parents, with perhaps the advice of the midwife or elders, but not by the government. Besides physical defects, reasons for killing were illegitimacy, adultery, and prostitution. In virtually all ancient cultures, as well as many contemporary primitive ones, infanticide has existed. Today, newspapers report cases of infanticide about once a week.

Aristotle described the pronatalist policy of Sparta, the frequent enemy of his home city of Athens, noting that its laws provided that ".... he who had three children should be excused the night watch, and that he who had four should pay no taxes."[10] Unmarried men suffered indignities. The Spartan war machine wanted the maximum number of soldiers.[11]

In Rome the emperor Caesar Augustus ordered all bachelors to marry. The patricians worried that their numbers were declining. The pronatalist legislation offered patricians rewards for having children, bestowed on mothers the honor of wearing distinctive clothing, prevented bachelors from claiming their inheritance, gave subsidies for children, criminalized abortion, and made infanticide a capital offence. Eventually the emperor extended the policy to all citizens.[12] The laws lasted for three centuries until Constantine rescinded them. As the first Christian emperor, his policy coincided with the growth of church monasteries and the emerging custom of celibacy for secular priests. Furthermore, many Christians of the period believed that the earth was full and that the maximum number of redeemed souls already existed. Because the Messiah had come, there was no need to continue procreation.[13]

Disease has played a major role in limiting population since the beginning of civilization. Archeological digs from ancient ruins show that living in cities increased disease. Smallpox, dysentery, typhus, measles, and bubonic plague caused the worst epidemics. Thucydides described the plague that afflicted Athens in 429 B.C. during the Peloponnesian War, killing a large part of the army and many civilians. The fall of the Roman Empire coincided with a population decline of 50 percent. The number of people living in Europe went from 44 million in A.D. 200 to 22 million in 600.

Bubonic plague struck Constantinople in A.D. 542, killing 70,000 in two years, then spreading to Italy and France. Bubonic plague is still found today, primarily in underdeveloped countries, with 1,000 to 3,000 cases a year. The last American epidemic was in 1925 in Los Angeles; about 10 to 15 cases occur each year, chiefly in the Four Corners area of the southwest. The worst plague in Europe was the Black Death, which killed one-quarter to one-third of the population between 1347 and 1352. This epidemic plague originated in China and spread west. It attacked the Genoese colony of Caffa in Crimea on the Black Sea. The sailors, soldiers, and merchants sailing home infected Constantinople and Messina before arriving at Genoa. From there the plague spread north to France and England, then east to Germany and eventually Russia in only five years. In many villages, the survivors were too few or too weak to bury their dead. Farms were abandoned for lack of hands. Insecure survivors experimented with loose living, fanatical religious movements, literary creativity, and novel forms of dress.

The decline of the native population of North and South America after their discovery by Columbus was even more extreme than the European epidemics. The total numbers in the New World fell from 42 million to 13 million in a century. The cause was chiefly disease. When Columbus arrived in Santo Domingo, he observed that the population was dense, "like the countryside of Cordoba." Twenty years later only a few hundred survived. In Cuba the tax census showed 112,000 Indios in 1512, and none by the end of the century. Mexico had the largest population (6.3 million Indios in 1548). This declined to 1.9 million in 1580 and 1 million in 1605. In Peru the Spanish counted 1.3 million Incans in 1572, which fell to only 600,000 in 1620. Prior to European contact, the future United States was home to five million Indians, which fell to a mere 60,000. In Canada where 300,000 Indians were alive in 1600, the number declined by two-thirds two centuries later.[14]

The first of many epidemics was smallpox, which arrived with Columbus to devastate Santa Domingo and Puerto Rico before moving on to Mexico. It annihilated the Aztecs, including the Aztec emperor, then spread south through Guatemala, eventually infecting the Inca Empire prior to the arrival of Pizarro and his conquistadores. The second epidemic was measles, which afflicted the Caribbean, Mexico, and Central America. Later, influenza raged and smallpox reappeared. Epidemics thought to be typhus struck in 1545 and 1574. Tuberculosis and chicken pox also killed many. The epidemics were so virulent because the American Indians had never been exposed to these diseases and so had never formed antibodies that would fight the infections. After an illness the survivors would develop immunity, but many died first. The medical situation was not entirely asymmetrical. When 12,000 English sailors and soldiers laid siege to the South American port of Cartagena, 8,000 died of yellow fever in two months. The susceptibility of virgin populations to modern illnesses continues to be a problem. Measles struck Canadian Eskimos and Indians in 1952 with devastating effect. In 1968 measles caused the deaths of many members of the Yanomono Indian tribe who live in the Orinoco River valley of South America.[15] In three decades the number of Yanomani dropped from 20,000 to 10,000.

The natural fecundity of the European settlers in North America was well known. Families often had a dozen children, and the unoccupied land gave an outlet for the growing population. Young people could move west and settle on virgin territory. The economist Adam Smith observed that "in the British colonies in North America, it has been found that they double in twenty or five-and-twenty years. Nor in the present times is this increase principally owing to the continual importation of new inhabitants, but to the great multiplication of the species."[16]

The French Canadians provide an excellent case study of natural increase because they were isolated and kept good records. Only a few thousand colonists ever immigrated from France, and since the establishment of Quebec in 1608, the population has grown to 6½ million today. Genealogical study shows that virtually all French Canadians descend from only 3,380 original settlers. By 1680 the few thousand had grown to 10,000, and this number increased 12 times to become 134,000 by 1784. This increase of 2.4 percent annually was due almost entirely to natural increase. Women who came to New France married an average of two years younger than women who remained in old France. Remarriage was more common too. Mothers tended to have babies more frequently, every 25 months rather than the 29 months for those who remained. Life expectancy was longer—20 years compared to a brief 15 years for those who remained. Each mother bore an average of 6.3 children.[17]

Although once established the North American colonists reproduced rapidly, the very earliest British settlers suffered terribly. The first colony planted on Roanoke Island in 1585 disappeared with hardly a trace. In Jamestown, 66 of the original 104 settlers died in the first year of famine and fever. Of the 6,000 people who came to the settlement in its first 14 years, only 3,400 survived. In Plymouth, half of the 101 who came on the *Mayflower* died the first winter. The Plymouth hinterland was strangely empty as nearly all the Indians had died in the preceding years. But by 1700 the British colonies were holding steady in births over deaths, and by mid-century, they were increasing rapidly. Benjamin Franklin wrote in *Poor Richard's Almanac* in 1749 that census data for Massachusetts and New Jersey that he had examined showed increases of a sixth and a third respectively in only seven years.[18] These were areas without much immigration.

THEORIES AND POLICIES

European writers of the early modern period analyzed and recommended population policies for their home countries as well as on their colonies. Francesco Patrizzi of Siena wrote in 1518 that although a large population had advantages of wealth and military manpower, overpopulation was a detriment. Niccolò Machiavelli warned that too many people would lead to poverty and disease. His English contemporary, Thomas More, advocated strict regulation of population in his Utopia. If a family has too many children, the state should remove them and give them to a less fruitful couple.[19]

In France in 1576 Jean Bodin, the king's attorney and member of the Estates General, recommended a large population. This would increase political stability because a greater number of citizens would have multiple interests that would tend to balance each other, and because it would support a middle class. German writers of the late seventeenth century feared that the population was becoming too dense. English opinion varied. Simply estimating population change was difficult due to the lack of historical data. Some observers pointed to the fact that the old churches could hold much larger congregations as evidence that numbers had shrunk. After the Black Death, writers advocated more people, but during the Tudor period, opinion shifted to believe that the country was overpopulated, leading to colonization in North America and strict regulation of rural people under the Poor Law. The growth of mercantile economic theory in the 1700s again reversed the direction of policy, arguing in favor of a larger population in the home country.[20]

For the coming century, England, France, and other countries came under the sway of the economic theory later labeled mercantilism. The goal of mercantilism was to stimulate industry by means of barriers to importation of manufactured goods. A country should strive for self-sufficiency, import only raw materials, and export finished products. In particular it should discourage using gold and silver to pay for imports. The government implemented mercantilism with monopolies and privileges. Mercantile principles dictated that colonies should trade only with the mother country.

The theory of mercantilism developed in reaction to the negative example of Spain. The gold and silver of Mexico and South American was squandered on luxuries purchased from Italy and the Netherlands, doing no good in Spain itself. Reacting to the Spanish case, the French finance minister Jean Baptiste Colbert reordered the country's economic policy along the lines of mercantilism. Serving King Louis XIV from 1658 to his death in 1683, he reformed the medieval taxes, which were unfair and discouraged investment. He encouraged local manufacturing and raised the quality of their production. The government built roads and canals. Colbert sent recruiters abroad to find skilled craftsmen and forbade emigration from France.[21] The proscription against leaving even extended to Protestants, whom Louis XIV was persecuting. As a result, France gained new production, expanded its foreign trade, increased its prosperity, and integrated its economy.

English mercantilism developed more gradually. Tariffs protected domestic industries. The rigid power of the medieval guilds to control trade was reduced. Sumptuary laws discouraged luxuries. Navigation laws gave priority to British ships, forbade foreign sailors on the ships, and protected forests that supplied ship timbers and naval stores. The government announced fish days to assure that fishermen would have a market. Colonies were required to provide the mother country with raw materials.[22] Tobacco from Virginia was an exception because demand for the "noxious weed" was so high. The stockholders of the Virginia Company chartered in 1606 included 20 of the highest nobility. Not surprisingly it received a monopoly, and importation of Spanish tobacco was forbidden. In 1624 King James I dissolved the original company and declared himself in charge. He assured the royal monopoly by demanding that all ships bring their cargo to the official warehouse in London, but many Virginians illegally smuggled their tobacco to other ports.[23] With the exception of tobacco, the American colonists did not benefit. For example the Navigation Act restricted their trade with France, Spain, and the Caribbean.

Mercantilism is of interest, not only because of its tenets on population, but also because it is the first example of a national policy. Its appearance in Spain, France, and England coincided with the emergence of the nation-state as the dominant form of government in Europe. By definition, one cannot have a national policy without a nation. In all three of these countries, mercantilism was the creation of the monarchy. In that age of absolutism, the king was the key. A central feature of the policy was to enhance the wealth of the monarch himself. King Philip claimed the gold of America, King James I dismissed the stockholders of the Virginia Company, and King Louis XIV grew rich on increased taxes. Notwithstanding royal selfishness, mercantile policy brought prosperity to the common people and expanded the middle class.

Present-day views unfairly berate mercantilism for its concern with keeping gold and silver within the nation. In fact few of its early advocates proposed to do this merely to gain the bullion itself. Their real goal was to invest the gold and silver in domestic production. The more sophisticated version of bullionism today would be called a policy of import substitution, the principle that encourages domestic manufacturing in place of importation. The term "mercantilism" was not used by its advocates, but was the name Adam Smith bestowed in critiquing it in *The Wealth of Nations* in 1776. Smith deplored its monopolies, protectionism, absolutism, and corruption.

In general, mercantilism held that a large population was advantageous, assuming the workers were skilled and willing. While colonies appeared beneficial, a question remained whether emigration would sap the mother country. Sir Josiah Child, a member of Parliament and governor of the East India Company, argued that England was better off without the emigrants:

New England, as everyone knows, was originally inhabited, and has since successively been replenished, by a sort of people call Puritans, who could not conform to the ecclesiastical Laws of England.... Virginia and Barbados were first peopled by a sort of loose vagrant people, vicious and destitute of means to live at home....[24]

His contemporary, Sir Matthew Hale, ignored the economic side of the issue. He observed that natural populations such as animals, birds, and fishes have a greater natural capacity for reproduction than man, but their numbers remain relatively stable. In contrast, mankind "will continually increase in a kind of Geometrical Progression...."[25] His terminology presaged Malthus. John Locke, the great philosopher of the future United States, held thoroughly mercantilist views on population, believing that a large population was beneficial, and that labor put the value on goods.[26]

By the middle of the eighteenth century, the European consensus was optimistic about the benefits of a large population, which would increase prosperity and provide soldiers for defense. Indeed, national wealth without a large population appeared to be an invitation to military invasion. Daniel Defoe glorified the agglomeration of people in London because it stimulated trade. Others wrote of the benefits of specialized labor due to density, and of throwing off the oppression of rural landlords.[27] Not everyone agreed with the optimists that urbanization would support endless increases. The Scottish economist James Steuart made the point that the number of animals is restricted by their food, and that this was also true for man. The French physiocrat, Victor Riqueti, Marquis de Mirabeau, believed that humans were near the limits of their subsistence.[28]

MALTHUS AND HIS ESSAY ON POPULATION

Thomas Robert Malthus came from wealthy rural family and was tutored at home until he entered Cambridge University in 1784, receiving a BA and an MA. He was ordained in the Church of England and served briefly in a parish. He married at the age of 38 and had three children (contrary to the canard that he had 11). From 1805 until his death, he was professor of political economy at the college of the East India Company at Haileybury near London. In 1798 he published his *Essay on Population.*

His stimulus was two-fold. First was debate in Parliament and throughout the country about amending the Poor Law. Clearly the program was ineffective. The number of poor was increasing, and the law had perverse incentives. His second stimulus was to counter the optimistic opinions expressed by economists under the influence of Mercier, Condorcet, Saint-Simon, and others who believed in the perfectibility of man. They maintained that enough food could be produced easily for the growing population. To Malthus, this was fantasy. He had only to look about to see the poverty caused by too many children. People, he argued, would reproduce until they starved. The life of the ordinary person was a desperate struggle against privation, disease, filth, and ignorance. Happiness was impossible because population always would outrun its food supply. He argued that the number of people would progress geometrically (2, 4, 8, 16, 32), but that the food supply would only increase arithmetically (2, 4, 6, 8, 10).

Malthus defined two types of checks on population: preventive and positive. Preventive checks limited births via moral restraint, vice, and birth control. The upper class successfully used moral restraint, but the lower class was incapable of this, so it used vice and birth control, which unfortunately were not enough. To Malthus, vice was masturbation, homosexuality, adultery, and prostitution, while birth control was coitus interruptus. The result was that the poor inevitably had too many babies. The second sort of checks—causing death—which he called positive, were famine, misery, plague, and war. He considered charity and paternalism, such as the poor laws, flawed as ways to curb population increase. Malthus believed that the upper classes were far superior and worried that trying to help the lower classes would divert resources from those who could advance civilization.

The publication was an instant success, immediately persuading many readers of its correctness and countering the dominant belief that a large population was desirable. The *Essay on Population* went through six editions, gaining facts and examples until the massive edition of 1826. Yet in spite of its expansion, it never had a solid statistical basis. Malthus's assertion that population increased geometrically while food increased only arithmetically did not rest on a firm empirical basis. His best data were a few reports on the increase of the North American colonists.

When Malthus observed the poverty of the peasants living in England, he also observed the wealth of their masters, the rich landlords living in their country estates. They passed their time hunting, causing Parliament to indulge them by enacting laws protecting the game from the peasants. Poachers could be imprisoned, whipped, or transported to Australia. Malthus applauded the prosperity of the landowners, urging them to spend their money on servants, thus providing employment. He extended his economic glorification of the rural aristocracy to its political benefits, which he believed guaranteed personal freedom. The landowner, however, did have a duty to be moderate in his desires and act paternalistically toward his tenants and servants.[29] At the same time, Malthus was an active member of the Whig Party, who advocated universal schooling, extension of the voting franchise, free medical care for the poor, and an end to child labor in factories.

Malthus's influence spread widely, into biology, economics, and politics. In his autobiography, Charles Darwin described how reading Malthus opened his eyes to the struggle for survival. Animals and plants breed prolifically (geometrically in the terminology he borrowed from Malthus). They are kept in check not by limited births, but by food supply and predation. Each individual competes with others of the same and different species, which results in survival of the fittest.[30]

Writing in 1844, the Communist Friedrich Engels took Malthus to task for lack of evidence. "Where has it been proved that the productivity of the land increases in arithmetical progression?" Engels countered that labor power can increase agricultural productivity, and that scientific progress is limitless and can increase at least as rapidly as population. Finally he noted that only a third of the total arable land on the earth was settled. The Valley of the Mississippi contained enough extra land to feed the entire population of Europe.[31]

Karl Marx also was a bitter enemy of Malthus's ideas, mentioning him 62 times in *Capital*. He denounced the *Essay on Population,* saying that it was "nothing more than a schoolboyish, superficial plagiary.... and does not contain a single sentence thought out by himself."[32] Marx's hostility stemmed from his recognition that Malthus's theory contradicted his own. Marx argued that the surplus product of labor, unfairly appropriated by capitalists, should be returned to the workers who created it in the first place. This would eliminate the cause of poverty. But if Malthus were right, the workers would breed until they ate up the surplus, and none would be left. Marx held that the problem only occurred under capitalism, and that the answer was a revolution to establish communism.

MIGRATION

Migration serves as both an outlet for excess population and a source of population growth. The phenomenon is older than history, and indeed older than the species itself. Genetically modern humans migrated out of Africa 1½ million years ago, first to the Middle East, and then farther east. They arrived in Europe half a million years ago. Lower sea levels due to glaciation locking up water permitted migration to Australia 50,000 years ago. To the north the lower seas permitted migration from Siberia to Alaska across the Bering land bridge.

The earliest written histories tell of invasions and migrations. The Hyksos invaded Egypt in the twenty-first century B.C., and the Philistines invaded Palestine in the thirteenth century B.C. From the eleventh to the seventh centuries, Greek tribes from the north invaded the present-day country, killing and driving out the natives. The Bible tells of the Hebrew exodus from Egypt, wandering in the desert, and the invasion of Canaan. Later

King Nebuchadnezzar conquered Judea, exiling thousands of the elite to Babylon, where Daniel wrote his apocalyptic vision of their return. Julius Caesar fought the Gallic War against the Helvetians, a Celtic tribe of 250,000, who wanted to migrate into Gaul to find better farmland and escape the pressure of German tribes from the north.

The great Atlantic migration started, of course, with Columbus. Settlement began immediately in Cuba, but only in small numbers. The conquests of Mexico and Peru were primarily military. Only after many years did the Spanish government encourage emigration of its own populace from the Iberian Peninsula, and even then the numbers were not large. The Portuguese colonized Brazil, but not in large numbers either. The French were even less fond of emigration. Only a few thousand left Europe in any one year, going to Canada and the West Indies.

The British, on the other hand, encouraged emigration to the New World. They were late in claiming colonies, nearly a century behind the Spanish and Portuguese. Jamestown began in 1607 and Plymouth began in 1620. A key difference was that the British, unlike other Europeans, settled their colonies with farmers. Their 13 North American colonies had the advantages of temperate climate, good rainfall, and fertile soil. With farm land obtainable for merely chopping down the trees, the North Americans prospered and multiplied.

In its first four decades of independence, the new United States of America received only 250,000 immigrants. At first, the turmoil of the Revolutionary War discouraged many people. Then in Europe the French Revolution and the Napoleonic wars disrupted travel and left many too impoverished to pay for their passage. Worse, the European governments on both sides conscripted soldiers to fight. Once the United States entered hostilities in the War of 1812, the stream again dried up completely. Nearly all American leaders eagerly welcomed immigrants. George Washington wrote, "The bosom of America is open to receive not only the Opulent and respectable Stranger, but the oppressed and persecuted of all Nations and Religions."[33] Congress passed a Naturalization Act in 1790 requiring only a two-year waiting period for citizenship. With the return of world peace in 1815, immigration resumed. At first the pace was moderate. Only 150,000 arrived in the decade of the 1820s, but the following decade it grew to 600,000. During a single decade of 1845–54, three million arrived. In proportion to an existing population of twenty million, this was the greatest wave ever.[34]

Many of the newer immigrants were Roman Catholics, and this produced a backlash. The original settlers of New England had long feared popery from French Canada or King James II. During the years leading up to the Revolution, the New England churches had led opposition to the British crown. In the nineteenth century, many Americans believed that Catholic support for the despotic monarchies of Europe was a threat. As Irish and German Catholics arrived, they established dioceses and built churches, which caused the old-stock Americans to feel endangered. Catholics opposed the generic religious instruction in the public schools. In 1841 the bishop of New York ran a Catholic slate in the city election. Protestants denounced this as a violation of the separation of church and state. When in 1853 the pope sent a controversial nuncio to the United States to negotiate internal church problems in Buffalo and Philadelphia, Americans publicized the fact that the monsignor had played a leading role in suppressing the democratic Italian Revolution in 1848.[35]

Those opposed to immigration, especially of Catholics, organized the secret Order of the Star Spangled Banner in 1849. To maintain their secrecy, when asked about their organization, they would say, "I know nothing." Soon they dropped their secrecy and entered active politics, adopting the name of the American Party. Their opponents, however, labeled them the Know Nothing Party. The party advocated restricting immigration and a 21-year residency requirement before becoming a citizen. In 1852 the party won many state and local offices and, in the 1854 election, won 43 seats in the U.S. House of Representatives. That was the high point, for the bitter debate over slavery overwhelmed the party, and it disappeared after 1860. It did leave one permanent legacy: the invention of the presidential nominating convention.

After the Civil War, immigration flooded the country. Twenty-five million came from then until 1914 when the outbreak of World War I terminated the flow. In the period from the Civil War to 1890 most came from northern Europe: the British Isles, Germany, and Scandinavia. From 1890 to 1914 the originating countries shifted from northern to southern Europe. Italy sent a total of more than five million. Mennonites and Jews fled from Imperial Russia.

Chinese immigration to the United States amounted to 300,000 in the years between the California Gold Rush and the Exclusion Act of 1882. Beginning in 1848 the Taiping Rebellion caused economic depression and famine. Its leader, Hung Hsiu-ch'üan, believed he was the son of God and the younger brother of Jesus Christ and predicted the end of the world. In California, the Southern Pacific Railway recruited the immigrants to build the transcontinental railroad. Whites there considered the Chinese a threat because they would work for low wages. In response Congress passed the Exclusion Act of 1882 barring their immigration for a period of 10 years and eventually made the ban permanent.

Soon after, Japanese immigration became a problem. Thousands entered San Francisco every month, stimulating nativistic hostility. The city school board imposed segregation. Whites feared they would lose their jobs to the Japanese laborers. Farmers feared the competition of Japanese farmers. In 1907 President Theodore Roosevelt negotiated a personal agreement (as opposed to a formal treaty) called the Gentlemen's Agreement, stipulating that the emperor would not issue passports to laborers and farmers. In return the school district immediately rescinded its segregation plan.

The immigrants prospered in America, adding to the economy and the population. Soon they entered politics. The Irish were the first and most successful, building powerful machines in Boston, New York, and other northeastern cities. They usually entered the Democratic Party since the old-stock Americans were already in control of the Republican Party. To counter the Democrats, the Republicans courted later arrivals like the Italians and Jews.

While most Americans at the turn of the twentieth century favored open immigration, a few found fault. In the Midwest, the American Protective Association, with an anti-Catholic bent, grew during the economic depression of the 1890s. The Knights of Labor, an early trade union, worried about cheap labor competition. In New England, old-stock Yankees smarted at their loss of political power to the Irish. The social scientist Richmond May-Smith published evidence that immigration was an economic detriment. Francis Walker, president of the Massachusetts Institute of Technology, blamed the declining birth rate of the Yankees on immigrant competition. These beliefs led to founding the Immigration Restriction League in Boston. In particular, the league opposed newcomers from southern Europe.[36]

The American policy of open immigration changed abruptly in 1921 with a temporary quota, made permanent three years later in the Johnson Reed Act. The change in policy derived from World War I. Patriotic citizens had been urged to concentrate on "Americanism," and after the Armistice, this transferred into hostility toward non-Americans. Moreover, people genuinely feared a flood of refugees from the destruction of the war. The mood of isolationism that opposed the League of Nations also opposed immigration. The Russian Revolution in 1917 frightened people. In the red scare of 1919, Attorney General Mitchell Palmer arrested 3,000 alleged agitators for deportation. A series of violent labor union strikes were led by foreign-born workers, and an economic recession in 1920–21 caused unemployment.

The Johnson Reed Act of 1924 took a new tack. From its earliest days the United States had maintained open immigration, at least from Europe. Toward the end of the nineteenth century, it had placed a few restrictions such as not admitting people who had tuberculosis, were insane, or were bigamists. In 1921 Congress enacted a one-year restriction based on 3 percent of the number of foreign-born from each country according to the most recent available census. This tended to favor those from northern Europe. Three years later Congress passed the permanent Johnson Reed Act, setting a maximum of 150,000 immigrants per year and establishing nationality quotas according to the proportion in the U.S. population. The rationale was to continue the existing ethnic ratios. Western hemisphere countries were exempt, but virtually all Asians were barred. When he signed the bill, President Calvin Coolidge said, "America must be kept American."

In the decade of the 1930s, immigration slowed to a trickle. The national origin system restricted those not from the British Isles and northern Europe. The Great Depression meant few foreigners wanted to come, since no jobs were available, and American citizens did not want to compete with the immigrants for jobs. Indeed, in the Southwest, many Mexican immigrants returned home. As the decade ended, the rise of Fascism and Nazism generated a wave of refugees. Their numbers were small, and they tended to be highly educated. As the likelihood of war increased, the European armies drafted soldiers, further reducing potential emigrants.

With the Allied victory in 1945, millions of people were displaced, and many wanted to emigrate. They included survivors of Nazi concentration camps, forced laborers, and exiles from Soviet conquest in the Baltic. The Russians expelled German-speaking minorities. Czechs, Poles, and Hungarians who opposed the Communist takeover in 1946–48 fled west and tried to get into the United States. The national origin system severely restricted their admission, since nearly none were from the British Isles and Scandinavia, and few were from Germany. In 1948 Congress passed the Displaced Persons Act, admitting 400,000, and in 1953 it passed the Refugee Relief Act admitting 214,000 more, which cleared up the backlog. When the Hungarian Revolution occurred in 1956, an additional 200,000 refugees fled. President Eisenhower invited 30,000 of them to enter as an exception to the law.

Because the Johnson Reed Act only limited immigration from the Old World, many people arrived from within the New World. In 1942 the U.S. government organized recruitment from Mexico for agriculture and railroad maintenance work. Others came from the Caribbean. Jamaicans and other West Indians were eligible to enter under the British quota until 1952, and 100,000 came. Because the United States had annexed Puerto Rico, its people were already American citizens and free to enter the mainland, which they began to do in large numbers in the 1940s.

During the 1950s and into the 1960s, American immigration policy remained tight with the continuation of the national origin quotas. Presidents from Truman to Johnson criticized it as contrary to American ideals and an embarrassment in leadership of the Free World against the Communists. By 1965 anti-Communist paranoia had moderated, and the black Civil Rights Movement logically extended to toleration of immigrants. Furthermore, the economy faced a shortage of skilled workers. The 1964 election gave Lyndon Johnson a mandate and increased the number of liberal Democrats in Congress. The Immigration and Nationality Act of 1965 revoked the national origin quotas, combining all slots from the Eastern Hemisphere into a single pool of 170,000 based on family unification (74 percent), work skills (20 percent), or refugee status (6 percent). The Western Hemisphere was subject to limits for the first time: 120,000 on a first-come first-served basis. Ten years later, the three categories were extended to cover the Western Hemisphere as well.

In spite of good intentions, the new law did not prove successful. After absorbing a backlog from Eastern and Southern Europe, the composition shifted almost entirely to the third world. Admitting so many skilled workers generated a brain drain from underdeveloped countries. Exemptions greatly exceeded the overall limit of 290,000. Asian immigration increased five times over. Most serious, illegal immigration grew greatly. Most immigrants settled in six states: California, New York, Florida, Texas, Illinois, and New Jersey. The percentage of foreign-born in the population rose from 5 in 1960 to 8 in 1990. By 1985 three cities—Miami, Los Angeles, and New York—had populations with more than 25 percent foreign-born. Family unification led to pyramiding; for example, an immigrant would be eligible to bring in his or her brothers and sisters. In turn, they could bring in wives, husbands, and children. The wife could then bring her family. Remarkably, public opposition to immigration was negligible.

The United States gave special privileges to political refugees. The 1948 Displaced Persons Act and the 1953 Refugee Relief Act admitted 620,000 in the aftermath of World War II. Focus then shifted to those fleeing communism. Eisenhower admitted 30,000 Hungarian freedom fighters on parole, a nebulous authority not based on a specific law. The 1957 Refugee Escape Act and the 1965 law limited eligibility to refugees from communism. After the fall of Saigon to the North Vietnamese in 1975, a total of 800,000 immigrated from Indo-China. In 1980 Congress passed the Refugee Act that set a quota of 50,000, abolished the parole power, and extended the definition to include more than flight from communism. The Reagan administration implemented the act selectively. Almost any refugee from a Communist regime was admitted, but almost none from right-wing dictators were admitted. In particular it turned back many from El Salvador and from Haiti.[37]

By the 1980s the problem of illegal immigration became increasingly serious. The official figure was six million living in the United States, and unofficial estimates were higher. These were overwhelmingly Mexicans and others from Latin America. In 1983 the government deported 1,250,000 "illegals" and estimated that it did not catch an equal number. Employers tended to favor the situation because the illegal immigrants were willing to take menial jobs for low pay, and if they caused a problem, they could be turned in to the Immigration and Naturalization Service for deportation. American workers, on the other hand, believed that the "illegals" were taking jobs away from them. Enforcement was difficult because the Immigration Service could take no action against employers. The 1986 Simpson Rodino Act was supposed to solve these problems. It granted amnesty to those who had entered the country prior to 1982, set penalties against employers who hired illegal immigrants, and authorized more money for enforcement. The law was almost a total failure. Many illegals were afraid to apply for amnesty, employers evaded penalties by relying on fake documents, and the money for enforcement was insufficient.

The situation took a turn toward the past by the end of the decade. The "new Irish" arrived, pushed out by economic recession in the Emerald Isle. Numbering as high as 150,000, they did not have the preferential quotas of the pre-1965 system, hence came illegally, and arrived after 1982, therefore could not qualify for the amnesty. In 1988 the Kennedy Donnelly Act provided for a lottery for 15,000 additional visas. The 1990 Immigration Act increased the annual employment-based immigration to 140,000, provided quotas for "underrepresented" countries like Ireland, Italy, and Poland, and raised the overall limit to 700,000. The Illegal Immigration Reform and Immigrant Responsibility Act, passed in 1996, further tinkered with the program, making it easier to deport criminals, reducing court intervention in Immigration Service proceedings, and building a 14-mile-long triple fence south of San Diego.

The Simpson Rodino Act recognized the impossibility of controlling immigration solely at the border. It required employers to obtain proof of citizenship, usually a social security card, or a work visa, known as a green card. In practice, this merely resulted in forgery. Unlike other industrial countries, the United States does not have identity cards. In an attempt to substitute, an amendment demanded that the states improve their drivers licenses by displaying the social security number, but when this led to financial scams, another amendment rescinded the provision. It did require the states to make their licenses tamper proof.

In 2004 President Bush proposed to reform the immigration law once more. His centerpiece was a Temporary Worker Program with three-year visas. This would help farmers and other employers. The proposal ran into immediate objection from two sides. Pro-immigration groups said it did not do enough and was merely a way to allow cheap labor for business. Anti-immigration groups said it was too generous, especially to those who had come in illegally. They sought strict enforcement of the existing law. As a bill moved slowly through Congress, an unusual coalition was forged. The National Council of La Raza, the Conference of Catholic Bishops, the National Restaurant Association, and the U.S. Chamber of Commerce worked together for its passage. Top leaders of the AFL-CIO also favored it. On the other hand, many ordinary workers and citizens opposed it. The best organized opposing group was FAIR, the Federation for American Immigration Reform. Another was Numbers USA. U.S. Representative Tom Tancredo was outspoken, denouncing "the cult of multiculturalism" and objecting to America becoming a "Tower of Babel."[38]

The Bush team appeared surprised and distressed by the reaction to what seemed to them to be a practical compromise. Many people only wanted enforcement. They wanted a fence to keep out Mexicans. Immigrant demonstrations began in February 2006. On April 10 immigrants in a hundred cities from Los Angeles to Boston took to the streets in protest. The crowds were as large as 100,000 in several places. They demanded rights for illegal aliens. Early protesters carried Mexican flags prominently, but this created such a backlash that they immediately switched to American flags. A few had even carried pictures of the Cuban revolutionary Ché Guevara.

Along the border with Mexico ordinary citizens have organized the Minutemen, a quasi-military organization that chases Mexicans illegally sneaking across. They lack the authority to actually arrest the illegals, but they alert the U.S. Border Patrol. The president announced plans to send 6,000 National Guard troops to the border. In Congress the most popular feature of President Bush's bill was a fence for seven hundred miles along the border. The boundary with Canada has long been lauded at "the world's longest undefended border," but that is becoming less true. The government announced it would require a passport or special identity card for Canadians entering and Americans returning.

Another dimension to the immigration debate is terrorism. After the September 11 attacks, everyone became concerned about keeping terrorists out of the country, and this spilled over into generalized antiforeigner attitudes. President Bush argued that his immigration reform bill would screen out dangerous terrorists.

Recent immigrants have contributed disproportionately to U.S. population growth, both through their own numbers and through a higher birth rate. Using Census Bureau data, Leon Bouvier calculated that without immigration after 1950, the U.S. population in 2000 would have been 232 million, 43 million less than the actual figure of 275 million. Immigrants amount to 35 percent of the growth since 1950. In the past 20 years, foreign-born women have been far more fertile than natives. In 1950, foreign-born women had a fertility rate 7 percent lower than native-born women, but in 1996 their rate was 34 percent higher. The result, combining immigration with the higher birth rate, meant that immigration contributed 61 percent of the total growth.[39] In a long-range analysis, Census Bureau demographer Gibson Campbell calculated the impact since 1790. He found that the 4 million inhabitants at the time of the first census produced 122 million Americans, and the total of 47 million immigrants over a period of two hundred years produced 127 million Americans, each category generating half.[40]

The foreign-born population of the United States is currently 34 million, or 12 percent of the total. This compares with 14 percent in 1900. Immigrants and their children are changing the racial and ethnic composition of the country. In 1900 Americans were about 90 percent white and 10 percent black. This was a decline for African Americans from the 20 percent measured in the first census of 1790. In 1900 there were 237,000 American Indians and 114,000 Asians. Today non–Hispanic whites constitute 72 percent, and blacks and Hispanics make up 12 percent each. Asians and Pacific Islanders make up the remaining 4 percent. The composition among the immigrants is 43 percent Hispanic, 26 percent white, 25 percent Asian and Pacific Islander, and 7 percent black. The Census Bureau projects that these minority groups will increase from the present 28 percent to 47 percent by 2050. Hispanics have overtaken blacks to be the largest group. Much of the growth is due to higher fertility rates for immigrants. Currently racial and ethnic minorities are contributing 40 percent of all births. The rate total fertility rate for Hispanics is 2.9, compared to 1.8 for non-Hispanic whites.[41]

In the late 1990s the Wilderness Society and Earth First! publicly opposed immigration. In 1998 the Sierra Club asked its members to vote whether the group should officially take a stand. Alan Kuper, who led the movement, argued that "slowing U.S. population growth from all sources (including immigration) will yield environmental benefits over the long term." Opponents countered that the policy was elitist. One said that "anti-immigration and population proposals draw attention away from the real factors responsible for environmental degradation: corporate greed, militarism, inappropriate technologies and the ever-growing gap between the richest and poorest segments of society." After extensive publicity, 550,000 members voted—60 percent opposed to restrictions and 40 percent in favor.[42]

U.S. FOREIGN POLICY

The American government had little reason to be concerned about the population of foreign countries prior to the medical revolution of the mid-twentieth century, when births began to greatly exceed deaths in the third world. In 1945 the population of China was in the range of 400 million, about three times the American population of 140 million. India had about 335 million.

Early attention came from John D. Rockefeller III, the brother of the future New York governor and U.S. vice president. His trip to the Far East in 1948 brought the problem to his notice, and he recognized that while the Rockefeller Foundation was working to improve health and agriculture overseas, it was not paying attention to its consequences in increasing population. At his suggestion, the National Academy of Sciences convened a conference of experts in health, agriculture, nutrition, and demography in Williamsburg. The result was to establish the Population Council, with Rockefeller as chairman of the board.[43] The council sponsored visits by experts to assist India, Pakistan, and other countries. Later the Ford Foundation donated funds to the council. The Planned Parenthood Federation, organized in 1939, supported the council, but its main thrust was within the United States.[44]

The United Nations had paid some attention to the issue. In 1946 it established a small population division to gather statistics. In 1952 it forecast a 1980 world population of 3.6 billion, which seemed high. In fact the actual figure turned out to be 4.5 billion. The UN convened the first World Population Conference in 1954 in Rome with six hundred participants from 40 countries. The combined opposition of Catholic and Communist countries voted down its recommendations, and the population division was demoted to become a branch. The UN specialized agency, the World Health Organization, gave medical advice to India on birth control, but encountered opposition from Catholic countries.

By the early 1950s the Marshall Plan's use of economic aid appeared so successful in restoring prosperity in Europe and blocking communism that the United States extended economic aid to underdeveloped countries. It was now obvious that military aid and alliances such as NATO needed an economic side. In 1958 President Eisenhower appointed a committee to review the military and economic aid programs, naming William Draper as chairman. The committee consulted with experts and visited Japan, Taiwan, and Korea. While the Draper Committee originally had not intended to examine the issue of population, it soon recognized the problem. The report recommended that the United States give money for both planning and for birth control. When announced at a White House press conference, this drew criticism from the Catholic Church for being immoral, and President Eisenhower chose not to endorse it. He later explained that he personally was quite concerned with the problem, but was afraid to enter a controversy with the Catholic Church so close to the election.[45]

John F. Kennedy had the opposite problem. As the first Catholic to receive the presidential nomination since the disastrous defeat of Al Smith in 1928, Kennedy wanted to display his independence from the church hierarchy. In several well-publicized talks, such as before Protestant ministers in Houston, he declared that he did not consider Catholic doctrine to be binding on him for policy decisions. His strategy suffered a blow only two weeks before the election when bishops in Puerto Rico threatened to excommunicate anyone who voted for the popular governor, Luis Munoz Marin, because he supported public schools and birth control.

With the advent of the Kennedy administration, the State Department Policy Planning Council issued a report concluding that population control was "the largest single determinant" of economic development and recommended that the United States should sponsor research, disseminate information, and assist private organizations. It believed that direct programs were still too controversial. The secretary of state tried to suppress the report, but this attempt generated negative publicity when a member of Congress got hold of it.

Draper lobbied the Kennedy administration on birth control. In 1962 the president asked him to conduct a confidential investigation of the political and economic situation in Brazil. Draper returned with a very negative appraisal and told Kennedy that overpopulation in the northeastern part of the country was especially serious. The president's first reaction was that this was a problem the Ford Foundation should handle, but he came to realize that the government had to act. Within a year, the Agency for International Development authorized its overseas missions to serve as liaisons between foreign governments and private organizations like the Population Council, Planned Parenthood, and the Ford Foundation. AID added several population experts to its top staff. Kennedy publicly endorsed research on birth control, and Congress appropriated funds for the National Institutes of Health for research.[46]

In his 1965 State of the Union speech, President Johnson spoke of "the explosion in world population" and in an address to the United Nations said that "five dollars invested in population control is worth a hundred dollars invested in economic growth."[47] That year AID gave grants totaling $2 million to universities and to private organizations like the International Planned Parenthood Federation, the Population Council, the Latin American Center for Studies of Population and Family, and the Colombian Institute for Social Development. None of the money, however, was to be spent on contraceptives. For instance, AID refused to buy condoms, which the Indian government requested, but it did buy jeeps for local instructors to use to visit rural villages for lectures. The following year, faced with a famine, India requested wheat. Johnson made shipping the wheat contingent on the government's agreement to begin a massive birth control program.[48]

The program gained firm footing in the 1968 Foreign Aid Appropriation Act, with Congress appropriating $35 million. AID now had authority and funds to assist governments, UN agencies, private organizations, and universities. Contraceptives were legal now. When the Draper Committee made its report some years earlier, techniques were condoms, diaphragms, and douches. As the debate progressed during the 1960s, the Pill was introduced and became highly popular. Intrauterine devices were also introduced. At first the IUD appeared a panacea for underdeveloped countries. It was cheap and did not require the woman to keep track of her pills. Unfortunately, side effects were reported. Yet the Pill was expensive and required discipline.

Upon becoming president, Richard Nixon continued the existing AID programs, although he tried to reduce the overall budget. Six months into his term, he sent Congress a message on population, addressing both domestic and foreign issues. He called population growth "a world problem no one can ignore."[49] Nixon went on to praise both AID and the United Nations. The agency hoped that research could achieve a breakthrough, such as a once-a-month pill. AID transferred money to the National Institutes of Health, but NIH did not feel comfortable with applied rather than basic research. The conflicting missions of the two agencies stymied the program.

At the end of the 1960s, the informed public began to make a connection between population and environmental protection. For most of the decade, people had become more aware of both issues, but separately; now they converged. Ehrlich's book, *The Population Bomb,* was published in 1968. Ehrlich himself became president of a new organization known as Zero Population Growth. In 1969 two otherwise disparate groups meet jointly: the Association for Voluntary Sterilization and the National Conference on Conservation. The next year the National Congress on Optimum Population and Environment met. These joint meetings showed recognition that the two problems were interrelated.

The key group continued to be the International Planned Parenthood Federation, both in advocating and in delivering services. Draper continued his leadership role, chiefly by raising money. Its budget grew from $1 million in 1965 to $25 million in 1972. He doggedly lobbied Congress, government officials, and foundation presidents to expand programs. His contacts included both Democrats and Republicans. He traveled to all six continents, constantly promoting efforts to control population growth.

Abortion became a controversy during this period. The reason for its emergence on the political agenda was largely technological. Gynecologists developed a new method of vacuum aspiration that could be performed in the physician's office, reducing the expense and improving the convenience. Because state laws required that abortions be performed in a hospital, the laws needed to be amended. Seventeen states had removed restrictions by 1971. New York passed a law making abortion legal upon request, without the necessity to demonstrate a medical reason. This sparked opposition from Catholics, who believed it was immoral. They were joined by conservative Protestants, who did not object to contraception, but drew the line here. Abortion as a domestic issue spilled over into the foreign aid issue, drawing attention to the fact that abortion was a common method of birth control, for instance in Japan. Nixon announced that abortion was "an unacceptable form of population control" and that "unrestricted abortion policies, or abortion on demand, I cannot square with my personal belief in the sanctity of human life—including the life of the yet unborn."[50] His view, however, was not shared by the U.S. Supreme Court, which decided in *Roe v. Wade* that a woman had a constitutionally protected right to an abortion.

In 1974 the United Nations sponsored another World Population Conference, this time in Bucharest, Rumania. When the U.S. delegation proposed that developing countries establish targets for acceptable growth, these countries rejected the proposal and accused the United States of trying to subjugate them. They rebutted that the first world should aid the economic development of the third world. In the now famous words of the Indian delegation, "development is the best contraceptive."[51] This same year AID banned using its funds for abortions. The pendulum swung back with the Democratic Party victory of Jimmy Carter, who proposed doubling funds for international population programs in response to the recommendations of the *Global 2000* report. The 1978 AID appropriations law made aid depend on a recipient country's population program.

When the United Nations sponsored another conference in Mexico City in 1984, the Reagan administration reversed two tenets of American foreign aid. First, it declared that population growth was not a problem for development, and second, it announced that it would no longer support any private or UN agency that advocated abortions. This applied whether or not the agency actually performed abortions. It became known as the Mexico City Policy or, to its opponents, the global gag rule. In 1985 the United States withheld $10 million out of $46 million from the United Nations Population Fund. The rationale was that this was the proportion contributed to China, where coercive abortions were used. The next year, it discontinued all support to the UN program. The International Planned Parenthood Federation lost $20 million in government grants. The distinction between abortion and contraception was a nullity to many antiabortion interest groups such as the American Life League, Human Life International, the Pro-Life Action League, and Operation Rescue because they opposed all forms of birth control.[52] President George H. W. Bush continued the Reagan policy of not supporting any agency that included abortion among its options.

Within two days of his inauguration in 1993, President Clinton reversed the gag rule at home and abroad, announcing his decision on the twentieth anniversary of the *Roe v. Wade* decision legalizing abortion. The pendulum swung in response to voting blocks of the two parties. The Republicans get many votes from the religious right, and conservative Catholics and the Democrats get many votes from women active in liberal groups.

The Cairo Population Conference, which met in 1994, included 180 countries and gave a special place to nongovernmental organizations, with more than 4,000 groups attending. This followed the pattern of recent UN conferences like the Rio and Johannesburg earth summits. The conference took a different direction than those in Mexico City and Bucharest by de-emphasizing the economic consequences in favor of women's rights and needs. Three principles were health consequences, gender relationships, and social justice. The Cairo Program of Action did not support the traditional "neo-Malthusian" view of the need to control population first. This view had suffered from its identification with coercive policies such as the Indian sterilization and the Chinese one-child policy. The program spoke of the need for gender equality and affirmed women's rights to bodily integrity, informed consent, and sexual relations free of coercion. It advocated incorporating population issues within the context of reproductive health. The "neo-Malthusians" criticized the program on three points. Resources are limited, reproductive health is a broad and nebulous concept, and a medical approach may be beyond the capacity of third-world countries.[53] Ten years later, when it was time for another world population conference, none was held. Support was lacking at the United Nations as well as from the Bush administration.

CHINA

American foreign policy regarding population has often focused on the People's Republic of China, both because its growth has been so large, and because of its controversial one-child policy. With a population of 1.3 billion, it adds 8 million to the world total every year. Since 1979 the government has limited a family to only one child, a program enforced with abortion and sterilization. The United States has generally opposed this, more vigorously under conservative Republican administrations.

In fact, China is not the greatest contributor to world population increase. That place belongs to India, with just more than one billion people. Each year 15 million more Indians are born than die. The difference is due both to the birth rate and to the number of women of child-bearing age. China has 16 births per thousand, while India has 25 births per thousand. because China has more women of child-bearing age, the difference in the annual rate of growth is not so great. The total fertility rate is 1.8 for Chinese and 3.1 for Indians, meaning that China is well below the replacement level of 2.1 and India is far above it.

The population of China at the end of World War II is not known accurately. The Nationalist government estimated 475 million. Some believed it was as low as 350 million, an actual reduction due to war and famine. In ancient times, from the Han dynasty to the Song dynasty it fluctuated in the range of 40–50 million. The period from 1751 to 1851 was one of dramatic growth, from 207 million to 417 million, about 1 percent a year. Writing in 1798 (the same year as Malthus), Hong Liangzhi railed against the dangers of overpopulation.[54]

For thousands of years, the Confucian culture had emphasized that strong families were the natural order. Children and wives were to obey the husband. The eldest son was to inherit the property, which was held in common, and the younger sons were to obey him. Upon marriage, the bride would move to the husband's household, where she would obey her mother-in-law. Her chief function was to produce a son and heir. Daughters were held in low esteem, because they were expensive to raise and after marriage would no longer contribute to the family. Matchmakers arranged the marriages. Men often took second and third wives, and child brides were common. The individual families were integrated into the clan, which in turn were supposed to be subject to the emperor. This rigid system loosened after the 1911 revolution overthrew the emperor and the established the Republic. Western ideas penetrated, factories gave employment, young men and women moved to cities, and education improved. Legislation gave women a share of the family property. Moreover, among the poorest peasants, the lack of land undercut the economic basis of the family. During World War II, the government was eager for women to work in industry so men could serve in the army.

As soon as the war against Japan ended in 1945, the Communists under Mao Zedong resumed their civil war against the Nationalist army of Chang Kai Sheik, achieving victory in 1949. The next year they enacted a new marriage law, stating in its first article, "The feudal system, based on arbitrary and compulsory arrangements and the supremacy of man over woman, and in disregard for the interests of the children, is abolished." The man and woman, not a matchmaker, were to choose each other. Women had equal rights. The marriage ceremony, which was registered by the government, was to take place in front of a portrait of Mao Zedong. The government encouraged later marriages, recommending age 23–25 for women and 25–28 for men.[55]

At the time he came to power in 1949, Mao Zedong favored a large population. The Communists had needed manpower for their army and feared a counter attack by the Nationalists. The whole country had just expelled the Japanese after a decade of warfare. In the years right after the war, the country had suffered a severe famine, with millions dying. Mao rejected the idea that the cause was too many people. He wrote that:

It is a good thing that China has a large population. Even if China's population multiplies many times, she is fully capable of finding a solution: the solution is production. The absurd argument of western bourgeois economists, like Malthus, that increase of population cannot keep pace with increase in production was not only thoroughly refuted in theory by Marxists long ago, but has also been completely exploded by the realities in the Soviet Union and the Liberated areas of China after their revolutions.[56]

This policy shifted in 1955 when the Communist Party announced that "reproduction be appropriately restricted." The purposes were to improve maternal health and provide for education. The effort was modest due to a squeamishness of the party cadres to discuss sexual behavior, and a lack of contraceptives. As the Communists consolidated power, however, population control seemed a more appropriate function of the government. More than a million people were resettled in Mongolia and the northern and western frontiers in order both to strengthen their defense against the Soviet Union and to relieve crowding in the east. The growth of the cities seemed to be getting out of control. In three years, 840,000 people moved into Shanghai. Food had been short when the 1952 harvest was poor. In 1957 Mao Zedong announced a 10-year program for family planning.[57] Almost at once, the policy got lost with Mao's Great Leap Forward, a fantastical scheme for instant industrialization, including backyard steel making. Rural communes were supposed to improve farming. According to official propaganda, agricultural and industrial productivity would solve the problem of poverty. At the same time, China felt threatened militarily. The Soviets withdrew their friendship and ended economic aid. Tibet began an armed revolt, border warfare began with India, and the Nationalists seemed to threaten an invasion. The Chinese leaders feared that the United States and the Soviet Union would begin a suicidal nuclear war. A larger population seemed to offer at least some protection.

By 1963 the birth control program was back on track. The Great Leap Forward proved a great disaster, an estimated 30 million died, and the military threats passed. Food was scarce, and deaths exceeded births in at least one year. From 1966 to 1969 China suffered another convulsion of Communist direction with its Cultural Revolution. Mao Zedong suddenly decided that the country had lost its Marxist zeal and had fallen into bourgeois error. His remedy was to order professional cadres from the cities to move to rural areas where they could be reeducated. Millions were sent to the countryside. Part of the rejection of urban expertise was to create thousands of "barefoot doctors," who were supposed to care for the medical ailments with few professional qualifications. While overall the Cultural Revolution was another disaster, it did expose the experts to the poverty and backwardness of the countryside. Moreover the barefoot doctors were ideal for the low-level medical functions of running a birth control program.

China's return to normalcy after the Cultural Revolution saw renewed concern about population size. It no longer seemed so vulnerable militarily, and its growth rate was an astounding 2½ percent a year. The fourth five-year plan called for an urban growth rate of 1 percent by 1975. The fifth five-year plan continued the 1 percent rate and projected a reduction to one-half of one percent by 1985 and a zero rate by the end of the century. The methods were to be delayed marriages, greater space between children, and fewer children overall. Its slogan was "one is good, two is all right, and three is too many." Contraceptives were now to be distributed free, and women could get free hospital care for abortions, IUD insertions, and sterilizations. Propaganda attacked the Confucian beliefs about fertility, including the preference for sons.[58]

By the end of the 1970s, the policy was working. The birth rate had declined from 34 per thousand to 18 per thousand.[59] But this was not enough. Because of the baby boom of the 1950s, more young women were of child-bearing age. The total numbers were going up sharply. The only solution was one-child families. This required both positive and negative incentives. First, a couple who wanted to have a child needed to get permission. The rewards for complying were a monthly subsidy, priority in housing, free medical care, maternity leave, exemption from tuition, and additional vacation. Those who did not register their marriages, had a baby without permission, or had too many babies did not get these benefits and had to pay a fine.

The one-child policy was unpopular, particularly in rural areas. Parents wanted more than one, and they especially wanted sons. For a peasant, a son has a duty to care for his old parents, whereas a daughter does not. One consequence of the one-child policy was, of course, abortion, but when that did not occur, another option was to kill the unwanted babies. A second or third child might be smothered at birth or drowned in a pond. In cities, obstetricians would give the mother an injection that would cause a stillbirth. Hospitals were penalized if unauthorized babies were born. Because of the desire to have sons, parents might kill their baby daughters. This is, however, against the official government policy.[60] Because the government will make an exception and permit a second birth when the first child was deformed, some parents maim their daughters. The only child, the product of this policy, ended up smothered in attention from their parents and grandparents. If he were a boy, he gained the nickname of the Little Emperor. More sympathetic observers concluded these only children were not brats, but were thriving from the love of their families. Another consequence of aborting female fetuses is a shortage of women for marriage. The sex ratio is 113 males to 100 females. Chinese men are importing brides from Vietnam.

In recent years, the government has eased its program slightly in response to domestic and foreign opposition. A few localities have permitted a second child, but this does not have the sanction of the Beijing regime. Because parents have tested the sex of their fetuses in the uterus with ultrasound or chromosome tests in order to assure a boy, the practice has been made illegal. Ethnic minorities, like the Tibetans and the Uighurs, who comprise 8 percent of the population, are supposed to be exempt from the limits on the numbers of children, but in fact, the government demands that they use IUDs and abortions. In answer to foreign criticism, the government has soft-peddled its programs, denying that quotas are rigid or that women are forced to abort or be sterilized. Few critics accept these protestations. The tenth five-year plan, which was forced to revise the population projection to 1.3 billion, reiterates the one-child policy.

China's neighbors have varying situations. India, being a democracy, has not been effective in coercing birth control. An unpopular forced sterilization program in 1976 ended with Prime Minister Indira Gandhi's emergency dictatorship, and then the demise of her government. Vietnam experienced its own population explosion during the 1980s and hence adopted a two-child policy. At the end of the war in 1975 its population was 35 million, and now it is 80 million. On the other hand, prosperous countries in Asia have experienced a birth dearth, that is, a rate below replacement. Singapore, which is 80 percent ethnic Chinese, has a rate of 1.2 children per mother, about the lowest in the world. The government encourages Chinese to have more children and discourages the ethnic east Indians who live there from having too many. Japan, with a rate of 1.4 children per mother, is also worried about increasing its numbers.

Around the world, the population situation varies greatly. Nigeria, with 123 million people, is growing at a rate of 2.7 percent a year. Its total fertility rate is 5.7 children per woman. Ethiopia, with 64 million people, is growing at 2.8 percent a year and has a fertility rate of 7.1 children per woman. Latin America is growing more slowly than Africa. Brazil, the largest with 173 million people, is growing at only 1 percent a year and has a fertility rate of 2.1, comparable to the industrial countries. Mexico, with 100 million people, is growing at 1.5 percent a year and has a fertility rate of 2.7.

In contrast many countries in Europe face declining populations. Italy and Spain have rates of 1.3, well below the replacement level of 2.1. Moreover, these countries have high immigration from Africa and the Middle East, whose women account for many of the births. The rate for old-stock Italians and Spaniards is much lower. Russia has a rate of 1.3 children per woman and expects a population decline. This is a picture far different that Paul Ehrlich's population explosion.

AMERICAN NONPOLICY ON POPULATION

While U.S. foreign policy on population has had a high profile, there has been no self-conscious internal policy other than immigration. The political situation has been driven by medical technology, beginning a century ago. One woman, Margaret Sanger, epitomized the issues of birth control in the first half of the century. Sanger, who was the sixth of eleven children, blamed her mother's death at 49 on excessive child bearing. Born into an Irish Catholic family in Corning, New York, she attended college and nursing school. Sanger practiced obstetrical nursing on the Lower East Side of New York City, where she became distressed at the poverty, the number of unwanted babies, and the high mortality of both mothers and infants. The women lacked information about sexuality and could not get contraceptives. In 1912 Sanger shifted to providing birth control information and contraceptives full time. In addition to the problems of ignorance and poverty, she faced a governmental one: laws forbidding information and supplies because they were obscene. For years Sanger worked to change the laws and publicize contraception. She founded the Birth Control League, later named the Planned Parenthood Foundation. Churches, both Protestant and Catholic, opposed her early efforts, but by the 1950s, most Protestants no longer did. Catholic opposition continued, both to contraception and abortion.

In 1950 Sanger learned of medical research that eventually led to the birth control pill and backed it with money she and the Planned Parenthood Federation raised. Interestingly the chief scientist, Dr. John Rock, was a devote Catholic who attended mass every day. The Pill accomplished within a few years what political action was taking decades to achieve. The method was immediately popular with women, and in 1968 the Supreme Court ruled that information could be disseminated.

The contraception revolution sparked another revolution (one to reform laws restricting abortion). All states forbade the procedure except for medical reasons, typically to protect the life of the mother or to prevent the birth of a severely deformed baby. Rape was also a reason. The laws had been enacted at the turn of the century as medicine became more professional. Previously abortions had been largely the preserve of women—midwives, healers, relatives, and the pregnant women themselves. Physicians had persuaded state legislators that they should medicalize the field to save lives. Lay abortions were notoriously dangerous, and the introduction of antiseptic surgery during the nineteenth century was an improvement. The introduction of antibiotics in the 1940s greatly reduced the risk. Better medical care undermined the justification of saving the mother's life, so protecting her mental health became a rationale. By the 1960s physicians were firmly in control, and the decision making was formalized. Hospital review boards had to approve each abortion, putting impediments in the way.

Women resented losing control of their bodies to physicians, nearly always male. Other factors worked to change the laws. The contraceptive revolution with the Pill empowered women in the area of sexuality. Women were becoming better educated, and with less expectation of automatic motherhood, they entered other careers. The Civil Rights Movement in the South showed how social protest could change society. Technology improved so that a physician could perform the abortion in an office; hospitalization was not necessary.

In 1962 a pharmaceutical tragedy struck Europe. When a new sleeping pill, Thalidomide, went on sale, its horrific side effect was deformed babies. Several thousand infants in Britain, Germany, and Belgium were born with flippers in place of arms or legs. Most of the children were otherwise normal. In Phoenix, Arizona, a pregnant woman realized she had taken Thalidomide, which her husband had brought back from a trip to Europe. Her physician recommended an abortion, but no hospital would agree to perform the operation, and she had to fly to Sweden to abort the fetus. The widespread publicity brought attention to abortion restrictions. A second medical tragedy beginning that year was an epidemic of rubella: German measles. While rubella is a mild disease for the mother, in 30 percent of the cases it leaves the child blind, deaf, or mentally retarded. Most obstetricians recommended abortion in these cases, but they became worried that this might violate the law.

In Chicago a clergyman, a physician, and a lawyer founded the Illinois Citizens for the Medical Control of Abortion in 1966. It grew out of their efforts to change the state law that forbade social workers from distributing birth control information.[61] Three years later the Illinois group convened a national conference that established NARAL, the National Association for the Repeal of Abortion Laws, later changed to the National Abortion Rights Action League. Support came from feminists, physicians, social workers, lawyers, and liberal Protestants, especially Episcopalians, Methodists, and Presbyterians. While the Planned Parenthood Federation was not an early supporter, it did endorse the repeal strategy two years later. Hawaii in 1970 became the first state to repeal its abortion law. A few months later, by a narrow vote, New York also repealed its law. Other states, while not repealing their criminal laws, had reformed theirs so abortions became easier to obtain. Governor Ronald Reagan signed the California Therapeutic Abortion Act in 1967.

Shortly the floodgates opened. Women could either get an abortion at home or travel to another state like New York. Although the American political system pays lip service to the idea of federalism whereby the states maintain their autonomy (especially in criminal law), in fact the pressure for national uniformity is great. The Supreme Court followed medical science rather quickly in 1973 in *Roe v. Wade*. By a vote of seven to two, the Court ruled that a woman had the right to an abortion in the first trimester of pregnancy, contending that it was part of her "right to privacy."

In the immediate aftermath of the *Roe* decision, opposition came chiefly from the Catholic Church. The archbishops of New York, Philadelphia, and Washington directed all their priests to preach sermons against it. Bishops around the country denied the sacraments to women who spoke in favor of abortion, or even belonged to groups such as the National Organization of Women. Four cardinals testified before congressional committees, the first time a cardinal had ever done so. In 1973 the church spent $4 million lobbying Congress. It contributed most of the expense for the National Right to Life Committee until its blatant lobbying threatened its tax-exempt status. At the time the committee had virtually no other source of money.

As the 1976 presidential campaign heated up, abortion became an issue. By this time conservative Protestants were aroused. In the January Democratic caucuses in Iowa, Jimmy Carter garnered the most votes by appealing to conservatives within the party. Many were pro-life, that is, anti-abortion. Carter was a born-again Baptist who fudged his position. In the Republican Party, anti-abortion conservatives were much stronger. Governor Ronald Reagan challenged President Ford for the nomination. Even though he had signed the California Therapeutic Abortion Act, Reagan claimed to be staunchly pro-life. Jerry Ford, like Carter, fudged his position, saying he opposed abortion except when necessary to save the life of the woman or in the case of rape, and favored a constitutional amendment that would let the states decide. His outspoken wife, Betty, had earlier told the newspapers she was strongly in favor of the *Roe* decision. Although Reagan did not win the nomination that year, his conservative supporters got a plank in the Republican Party platform promising "a constitutional amendment to restore the protection of the right to life of unborn children."[62] In the end, however, the issue was not prominent in the fall campaign.

Pro-lifers met with success in Congress by blocking money to pay for abortions for mothers on welfare. Medicaid is a joint national-state program, and a few states had restricted funds for abortions. When in 1976

Representative Henry Hyde of Illinois offered an amendment to the annual appropriations bill forbidding abortions (with no exceptions) it passed overwhelmingly. The Senate added exceptions for rape and the mother's health, and it became law. In the following years, Congress continued to forbid Medicaid spending for abortions. President Carter did not support Democrats who opposed this provision. Many politicians viewed these Hyde Amendments, as they came to be called, as a middle-of-the road position. They argued that this approach did not go against *Roe,* but nevertheless stopped government payment for abortions. They claimed it did not take away any rights. Moreover, welfare was unpopular, so any way to cut its budget was good. During this period the voting strength of pro-lifers and conservatives increased. Protestant opposition to abortion also increased.

Ronald Reagan won the Republican nomination in 1980, and in his campaign he promised to oppose abortion by appointing Supreme Court justices who would overturn *Roe v. Wade* and lower-court judges who would oppose abortion. He said it would be a litmus test, and he was largely successful. He appointed three Supreme Court justices in his eight years in office, and although they never reversed the *Roe* decision, they did limit it. Eventually he named more than half of the lower-court judges, who tended to be quite conservative and anti-choice. The Reagan administration extended the restrictions of the Hyde Amendments beyond Medicaid to include federal workers and military personnel and their wives. It denied funds to private social service agencies that gave information about abortion. This meant ones that did not perform the procedure themselves, so that a woman who asked her social worker about the option could not be informed. This was known as the gag rule. At the Mexico City world population conference the Reagan administration announced the global gag rule that U.S. AID would not fund international groups under the same rubric. As at home, the government would not fund groups that gave information about abortions, even though they did not perform the procedure themselves.

After his election in 1988, George W. H. Bush continued the Reagan policies, and only three months after his inauguration an opportunity occurred to extend them. Many states had placed restrictions that made abortions more difficult, such as waiting 24 hours, testing to determine the viability of the fetus, consent of the father, parental notification, and so forth. When one case, *Webster v. Reproductive Health Services,* came before the Supreme Court, President Bush had his Justice Department argue in favor, but the decision was ambiguous, with three different opinions, and did not overturn *Roe v. Wade.*

Anti-abortion groups increased their efforts during the 1980s. The most important one continued to be the National Right to Life Committee, but the most extreme was Operation Rescue. Founded in Binghamton, New York, in 1986 by Randall Terry, the group attacked abortion clinics to rescue the fetuses. Its tactics were picket lines, confrontation, and forcing their way into clinics. When arrested, Terry and his partners would refuse to pay a fine, or even to identify themselves, in order to clog the court system (borrowing a tactic from the Civil Rights Movement). In 1988 Operation Rescue descended on clinics in Atlanta at the time of the Democratic National Convention. More than 1,300 demonstrators were jailed for trespassing. The group drew support chiefly from evangelical Protestants.[63]

Pro-choice groups cheered the election of Bill Clinton, who quickly ended the gag rule and reversed the Mexico City policy. On the negative side, Congress again passed the Hyde Amendment in the annual appropriations law. After only two years, the Republican Party achieved a majority in the House of Representatives for the first time in 40 years. Congress passed the Partial Birth Abortion Bill, which declared illegal the specific technique of dilation and extraction performed after the twentieth week of pregnancy. Pro-lifers argued that this procedure killed a fetus that was viable, while physicians rebutted that it was the safest method for a late-term abortion. It failed to become law when Clinton vetoed the bill.

The 1990s had witnessed incidents of violence against abortion clinics and physicians. Demonstrators like Operation Rescue often heckled, jeered, and tormented women trying to enter clinics. Activists set fires, threw acid, and exploded bombs. In 1993 a right to lifer assassinated David Gunn, a physician at his clinic in Florida. The next year an assassin killed another physician and his volunteer bodyguard. Later that year a gunman shot dead workers at two separate clinics in Boston. In 1998 a physician was killed by a sniper in his home near Buffalo. In reaction to these acts, Congress, with a Democratic majority still in the House in 1994, passed the Freedom of Access to Clinic Entrances Act, making interference a federal crime.

While campaigning in 2000, George W. Bush promised to oppose abortion. Two days after his inauguration on the anniversary of *Roe v. Wade,* he restored the global gag rule. He also sent a message to pro-life rally gathered in Washington that day declaring his solidarity with their cause. When Congress again passed the Partial Birth Bill, Bush signed it. Within hours a federal judge blocked it with an injunction because it violated *Roe* vs. *Wade,* sending it on a long appeal process. Otherwise, the Bush administration did not do much more than some symbolic statements. With the abortion issue so involved in the courts ever since *Roe,* attention centered on the views of nominees to the high court. At the hearings to recommend whether or not to approve a new justice, senators asked questions designed to elicit future opinions. The convention was that a nominee was not obliged to promise to vote for or against, but nevertheless he or she dropped hints that would encourage one side without alienating the other. For example when the chief justice nominee John Roberts testified in 2005, the Republicans tried to get him to agree to overturn *Roe* while the Democrats tried to get him to say he would uphold it. Roberts turned out to be a master of long, opaque answers couched in terms of "stare decisis," "respect for precedent," and "settled law."

For more than three decades, the issue of abortion has been one of the sharpest divisions between the two parties. Republicans put the anti-abortion plank in their platform in 1976, they have often called for a constitutional amendment to reverse *Roe v. Wade,* and they have promoted restrictions like the Partial Birth Law. Democrats have been staunchly pro-choice, meaning a woman should be free to make up her own mind. Indeed they have been hostile to members of their own party who oppose abortion, possibly costing themselves votes in elections. A few Democratic leaders have tried to counter this and welcome so-called pro-life Democrats.

Public attitudes vary geographically. The South Dakota legislature voted to ban all abortions, even in cases of rape, incest, and threat to the mother's health, a clear violation of *Roe v. Wade.* The objective of the Republican majority was to give an opportunity for the U.S. Supreme Court to have a case to use to reverse the 1973 case. Pro-choice citizens gathered enough signatures to put the law before the voters, where it was overturned 56 to 44 percent.

Abortion has been an element of the values agenda of the Republican Party. The strategy is to give the voters a consistent set of issues that will appeal to social conservatives, such as opposition to gay marriage and gun control and support of government prayer and flying the flag. George W. Bush promised during his campaigns to promote faith-based programs, that is, ones provided by churches and religious groups. He and other Republicans often tied their campaigning to evangelical Protestant leaders. They were often invited to the White House. In spite of the rhetoric, however, not many of these values were translated into laws and programs. The president did not push an amendment to reverse *Roe,* and the Office of Faith-Based Initiatives did little. A former staff member created a stir when he published a book telling that those closest to the president ridiculed and insulted the religious leaders behind their backs.

EUGENICS

While an observer at the end of the twentieth century would conclude that the United States had no self-conscious policy on population, he or she would have come to the opposite conclusion at the beginning of the century. A hundred years ago, many politicians, scientists, and statesmen actively advocated eugenics. The term was coined by Francis Galton, the famous British biologist and statistician, from the Greek word for "good in birth." It was, in essence, livestock breeding for humans. Galton's idea emerged in 1865 from the revolution in biological science sparked by the theory of evolution discovered simultaneously by Charles Darwin and Alfred Russel Wallace. In a study analyzing scientists belonging to the Royal Society, he found correlations with intelligence, longevity, health, madness, crime, and other variables. Additionally he found that distinguished families did not have enough children to replace themselves. Indeed, they had very few. The families that did have many descendants were at the bottom of the social scale: the poor, the uneducated, and the unhealthy. This trend troubled Galton.

Darwin himself had privately recognized the implications of his theory of evolution for human beings, but for a long time sought to avoid the topic because it would be too controversial. His co-discoverer, Wallace, was bolder, addressing the issue in an 1864 article that speculated that for humans, natural selection began to act on the brain so people became more advanced mentally and morally. He argued that selection favored societies whose members were rational and altruistic. In time the higher races would crowd out the lower ones. The result would be a bright paradise. Primitive people like the Indians of North and South America, the Australian aborigines, and the Africans would die out.[64] This was, of course, an accurate picture as of the nineteenth century of these people's mortality record since European contact.

Galton believed that society should address the problem by encouraging the most distinguished men and women to have more children. He concluded that charity and Christianity gave too much assistance to the unfit, who should not be protected from threats to their survival. Darwin, who happened to be Galton's cousin, accepted the negative implications of human evolution. Indeed Darwin's knowledge of livestock breeding had influenced his theory of natural selection. In the *Descent of Man,* he wrote, "There should be open competition for all men; and the most able should not be prevented by laws or customs from succeeding best and rearing the largest number of offspring."[65] Galton's most famous disciple, the statistician

Karl Pearson, maintained that progress depended on a harsh struggle among nations and races. Wallace, on the other hand, came to believe that negative eugenics would become an excuse for class bigotry and would postpone necessary political reform.

The late nineteenth-century British philosopher and scientist Herbert Spencer coined the term Social Darwinism, meaning that the process of natural selection would operate on human beings as well as animals. Life was a struggle for existence leading to the survival of the fittest. The best individuals would survive and reproduce. On a larger scale, nations and cultures also competed for survival. The theory gave support to laissez-faire capitalism, the class system, and foreign colonization. The innate superiority of certain individuals was demonstrated by their wealth; they had won the struggle, and the poor had lost. Spencer and his followers strongly opposed action by the government to interfere with the natural process.

American eugenicists studied the problem in the field. In upstate New York a prison reformer, Richard Dugdale, learned of the Juke family. The criminal careers of numerous Jukes extended over several generations. Besides their criminality, the family members were feebleminded and impoverished. In Pennsylvania and New Jersey, Henry Goddard chronicled the two branches of the Kallikak family, one from a feebleminded mother and one from an intelligent one. The bad branch included criminals, prostitutes, slackers, drunkards, and morons. Moreover, the bad branch had many more children than the good branch.[66] Dugdale and Goddard calculated the heavy burden on society of these unfit families.

During the nineteenth century, state governments had established asylums for the feebleminded as well as the mentally ill, the tubercular, and the poor, and as they expanded, their costs became burdensome. One benefit, however, was that the inmates, segregated by sex, were prevented from having children, thus reducing the long-term cost to society. Sterilization appeared to be the logical next step. Dr. Harry Sharp of the Indiana State Reformatory enthusiastically adopted the new technique of a vasectomy to sterilize five hundred inmates. In 1907 a state law authorized compulsory sterilization of confirmed "criminals, idiots, rapists, and imbeciles."[67]

The legality for sterilization was uncertain until the U.S. Supreme Court upheld a Virginia law authorizing sterilization in 1927, after which 30 states adopted similar laws. Eventually 60,000 men and women were sterilized. Justice Oliver Wendell Holmes wrote the majority opinion, stating:

It is better for all the world if, instead of waiting to execute degenerate offspring for crime or to let them starve for their imbecility, society can prevent those who are manifestly unfit from continuing their kind. The principle that sustains compulsory vaccination is broad enough to cover cutting the Fallopian tubes. Three generations of imbeciles are enough.[68]

Somewhat earlier, Theodore Roosevelt had written in favor of steriliza-tion: "I wish very much that the wrong people could be prevented entirely from breeding; and when the evil nature of these people is sufficiently fla-grant, this should be done. Criminals should be sterilized and feebleminded persons forbidden to leave offspring behind them."[69] He considered having large families to be natural, in contrast to the materialism and corruption that afflicted the nation. The natural instinct to mate and propagate would strengthen the country. Old-stock Americans were being outbred by immi-grants. Roosevelt advocated that these families should have six children. He opposed Margaret Sanger's birth control crusade because the upper classes were more likely to limit their babies than the lower classes. The American race (and he typically conflated race and nationality) had a duty to guide the people of the Philippines, just as the British race had a duty to guide the people of India.

As in Britain and the United States, people in Germany were inter-ested in eugenics. The Society for Race Hygiene, established in 1905, was alarmed about the low fertility of the professional classes. For its first three decades, race meant the human race or the white race. Only a few members promoted the Nordic race or the Aryan race. Indeed, a leading member, Alfred Ploetz, was quite positive about the merit of the Jewish race. On the other hand, Adolf Hitler zealously believed in the superiority of the Aryan race. Once in power, the Nazis coopted the Society for Race Hy-giene. They soon passed a law mandating sterilization for feeblemindedness, schizophrenia, genetic epilepsy, blindness, or deafness. Racial purity was still not grounds. The law authorized genetic health courts to be set up in every city and county. From 1934 to 1939, between 200,000 and 400,000 people were sterilized.[70] In 1939 the Nazis changed course. Sterilization would take too long; euthanasia was faster. Besides the mentally ill and feeble, and the blind and deaf, the Nazis added the racially unfit. The Nuremberg laws defined and degraded Jews. Other laws added gypsies, homosexuals, and Negroes. After the war was underway, the Nazis simply killed Jews, gypsies, and the physically deformed without bothering about the law or the health courts. The grand total for the Holocaust was six million.

After the defeat of the Nazis, virtually no one in the West had a good word to say about eugenics. Hitler had permanently discredited the movement. No one wanted to recall that, for example, Margaret Sanger had advocated birth control under the slogan of "fitter babies." She also wrote of breeding "a race of thoroughbreds." Sanger and others had also been advocates of free love, in other words, sexuality without the fetters of marriage or convention. Nazi atrocities were not the only nail in the coffin of eugenics. Psychologists shifted to believing that environment, not genetics, was the key. B. F. Skinner promoted the theory of behaviorism, meaning that behavior was a response to a stimulus from the environment. It was not instinctual, and certainly not genetic. This made it difficult to say who should breed and who should not.

Although the word eugenics is not uttered today, parents routinely test the genetic makeup of their future child. Every year American physicians perform 100,000 amniocentesis procedures, extracting the amniotic fluid surrounding the fetus of a pregnant mother to test for genetic flaws. It can identify 50 metabolic diseases as well as chromosomal anomalies and neural tube defects, such as spina bifida. Obstetricians perform 2.6 million sonograms annually to check for abnormalities such as Down syndrome, spina bifida, or cystic fibrosis. Two-thirds of pregnancies are checked. Genetic counselors advise on the likelihood of inheriting hemophilia, Tay Sachs, sickle cell anemia, muscular dystrophy, or others of the 3,500 disorders now linked to genes. Besides parents, the health insurance industry has a stake because it is cheaper if only healthy babies are born. Besides this genetic testing, 30,000 babies are born annually as a result of procedures like in-vitro fertilization.

In 2001 the use of embryonic stem cells for medical research and therapy brought the scientific establishment into conflict with anti-abortion groups like the National Right to Life Committee. These unspecialized cells, found in embryos only a few days old, are the precursors of all tissues of the body like the heart, nerves, skin, stomach, and muscles. Scientists harvest them from discarded embryos from in-vitro fertilization clinics. Experiments show they relieve symptoms of Parkinson's disease and diabetes. Research is in the earliest stages, but potentially stem cells could treat dozens of diseases. Right to lifers believe that the experiments will generate demand for embryos, which will result in thousands of abortions. Clinton supported stem cell research, but once Bush came into office, he felt the pressure of the anti-abortion groups.

After several months of highly publicized seeking advice, Bush announced his decision, which at first appeared good for both sides. Because some colonies of cells already existed in medical laboratories, scientists could continue to use them in research, but in order to prevent future abortions, no new colonies could be created. The president said that 64 strains were available.

This compromise was, of course, compared to King Solomon's proposal to cut the baby in half. Unfortunately, closer investigation showed that the actual number of usable strains was only a fraction of the supposed 64, and the result was to cripple medical research. The president was rather surprised to find that his fellow Republicans did not all line up behind him. Stem cell research brought out a cleavage with those who wanted the scientists to advance their research. Nancy Reagan, wife of the ex-president crippled by Alzheimer's disease, opposed Bush's policy. So did Senator Orrin Hatch, an extreme conservative, but a zealous advocate of medical progress.

President Bush controlled research using stem cells by means of the regulations and grants from the National Institutes of Health. NIH would only give permission and money to universities and institutions that agreed to abide with the conditions. Moreover a research center could not share laboratories and equipment. In a move unusual in the United States, where the national government has nearly all the funds, several states invested several billion dollars to pay for stem cell research. California, Massachusetts, and Wisconsin took the lead.

CONCLUSION

Writers like Paul Ehrlich have not been shy about framing the issue of population in apocalyptic terms, predicting starvation, disease, and war. Malthus forecast that the population would increase faster than the food supply, leading to starvation and misery. John Stuart Mill believed that poverty was inevitable because people followed their brute instincts to reproduce. Within the United States, immigration appeared to be dangerous as well as natural increase. The nineteenth-century Order of the Star Spangled Banner believed that immigrants from Europe, especially Catholics, threatened national values. Forty years later the Immigration Restriction League raised the same alarm, marshaling social science theories and data to support its argument. Today many Americans claim that high levels of immigration will tear apart society.

Within the United States, the dangers of domestic overpopulation have been too controversial for official commissions. Even in terms of overpopulation abroad, the issue has been controversial. When the Draper Committee reported the dangers in foreign countries to President Eisenhower, he wished it would go away. Many of the predictions have not come from official commissions, but from private groups like Planned Parenthood. Like the situation regarding the energy crisis, good statistics are crucial. In contrast to the United States, China has a comprehensive and successful policy of one child. It seems a poor model for the United States, however.

Regarding immigration, the United States did not even consider it to be a policy issue until after World War I, when the Red Scare and fear of European refugees brought it to the attention of Congress. In passing the Johnson Reed Act of 1924, Congress took a comprehensive view and arrived at a solution, which was that immigration should not upset the existence balance among the ethnic groups already in the nation. Although now discredited as bigoted, it was a good compromise for the wishes of the voters at the time. The 1965 Immigration and Nationality Act was supposed to meet the different attitudes toward civil rights, which it did, but it had at least one unplanned consequence, which was the flood it opened with its family reunification and skilled worker provisions. This was not the only factor, because this was a period of cheap airplanes and global communications. Again in 1986 Congress thought is had solved the immigration problem in a comprehensive fashion, but found the number of illegals soared. The bill President Bush sent to Congress in 2006 was supposedly comprehensive, but in fact had internal contradictions. On one hand it tried to keep immigrants out with fences and employer penalties, but at the same time permitted many to come as temporary workers.

NOTES

1. Paul R. Ehrlich, *The Population Bomb* (New York: Ballantine, 1970), p. 11.
2. Ibid., p. 33–34.
3. Ibid., p. 141.
4. Ibid., p. 147.
5. Ibid., pp. 164 and 173.
6. Paul R. Ehrlich and Anne H. Ehrlich, *The Population Explosion* (New York: Simon and Schuster, 1990), p. 9.
7. Ibid., pp. 190, 207, 209–213.
8. Genesis 1:28; 9:9 and 9:18; 15:5.
9. Massimo Livi-Bacci, *A Concise History of World Population,* 2nd ed. (London: Blackwell, 1997), pp. 22, 43, 38.
10. Aristotle, *Politics,* Book II Chapter 9, in *The Politics and the Constitution of Athens,* ed. Stephen Everson (New York: Cambridge University Press, 1996), p. 52.
11. E. P. Hutchinson, *The Population Debate* (Boston: Houghton Mifflin, 1967), p. 10.
12. Virginia D. Abernethy, *Population Politics* (New York: Plenum, 1993), p. 74.
13. John T. Noonan, *Contraception* (Cambridge, MA: Harvard, 1966), p. 83.

14. Livi-Bacci, *A Concise History,* pp. 55–56.

15. Ibid., pp. 59–60.

16. Adam Smith, *The Wealth of Nations* (London: Dent and Sons, 1964, originally 1776), vol. 1, p. 62.

17. Livi-Bacci, *A Concise History,* pp. 63–65.

18. James H. Cassedy, *Demography in Early America* (Cambridge, MA: Harvard University Press, 1969), pp. 160–161.

19. E. P. Hutchinson, *The Population Debate* (New York: Houghton Mifflin, 1967), pp. 16–17 and 3.

20. Ibid.

21. Joseph Spengler, *French Predecessors of Malthus* (New York: Octagon Books, 1965), p. 21.

22. Philip W. Buck, *The Politics of Mercantilism* (New York: Farrar, Straus and Giroux, 1975), pp. 14–17.

23. Donald A. Walker, "Virginia Tobacco during the Reigns of the Early Stuarts," in Lars Magnusson, *Mercantilist Economics* (Boston: Kluwer, 1993), pp. 144, 158, 160.

24. Quoted in Hutchinson, *The Population Debate,* pp. 51–52.

25. Quoted in ibid., p. 56.

26. John Locke, "Lowering of Interest" and "Civil Government," quoted in ibid., pp. 61, 60.

27. Hutchinson, *The Population Debate,* p. 100.

28. Ibid., pp. 116, 118.

29. David Cannadine, "Conspicuous Consumption by the Landed Classes 1790–1830," in *Malthus and his Time,* ed. Michael Turner (New York: St. Martin's, 1986), pp. 102–103.

30. Charles Robert Darwin, *Origin of Species,* (New York: Modern Library, 1993; originally 1859), chapter 3.

31. Friedrich Engels, "Outlines of a Critique of Political Economy," reprinted in Thomas Robert Malthus *An Essay on the Principle of Population: Text, Sources and Background, Criticism,* ed. Philip Appleman (New York: W. W. Norton, 1976), p. 150.

32. Reprinted in ibid., pp. 159–160.

33. Quoted in Maldwyn Allen Jones, *American Immigration,* 2nd ed. (Chicago: University of Chicago Press, 1992), p. 67.

34. Ibid., p. 79.

35. Ibid., p. 129.

36. Ibid., pp. 222.

37. Ibid., p. 279.

38. Holly Bailey, "A Border War," *Newsweek,* April 3, 2006.

39. Leon Bouvier, "The Impact of Immigration on United States Population Size," *NPG Forum,* November 1999; Ed Lytwak, "A Tale of Two Futures: Changing Shares of U.S. Population Growth," *NPG Forum,* March 1999. For an earlier version, see Leon F. Bouvier and Lindsey Grant, *How Many Americans?* (San Francisco: Sierra Club Books, 1994), pp. 64–73.

40. Gibson Campbell, "The Contributions of Immigration to the Growth and Ethnic Diversity of the American Population," *Proceedings of the American Philosophical Society* 136 (1991): 157–175.

41. Martha Farnsworth Riche, "American's Diversity and Growth," *Population Bulletin* 55 (2000): 10, 15–17, 44.

42. Danielle Knight, "Environmentalists Reject Anti-immigration Move," *Inter Press Service,* April 27, 1998; Dick Schneider and Alan Kuper, "Why We Need a Comprehensive US Population Policy," *Sierra,* January 1998.

43. Peter Bachrach and Elihu Bergman, *Power and Choice: the Formulation of American Population Policy* (Lexington, MA: D.C. Heath, 1973), pp. 44–45.

44. Phyllis Tilson Piotrow, *World Population Crisis* (New York: Praeger, 1973), pp. 13–16.

45. Ibid., pp. 36–40.

46. Ibid., pp. 63–64, 73, 74.

47. Lyndon B. Johnson, "State of the Union," January 4, 1965, *Public Papers of the President 1965* (National Archives, Office of the Federal Register), vol. 1, p. 4; Piotrow, *World Population Crisis,* pp. 89–90.

48. Jacqueline Kasun, *The War Against Population* (San Francisco: Ignatius Press, 1988), p. 86; Joseph Califano, *Governing America* (New York: Simon and Schuster, 1980), p. 52.

49. Piotrow, *World Population Crisis,* p. 169; also see Richard M. Nixon, "United States Foreign Policy for the 1970s," Message to Congress, *Weekly Compilation of Presidential Documents,* February 14, 1972, pp. 402–403.

50. *New York Times,* April 4, 1971, and May 7, 1972; Bachrach and Bergman, *Power and Choice,* p. 28.

51. Jocelyn DeJong, "The Role and Limitations of the Cairo International Conference on Population and Development," *Social Science and Medicine* 51 (2000): 943.

52. Sharon L. Camp, "The Politics of U.S. Population Assistance," in *Beyond the Numbers,* ed. Laurie Ann Mazur (Washington, DC: Island Press, 1994), pp. 124, 125, and 127.

53. DeJong, "Cairo," pp. 941, 942, 947.

54. Quoted in Penny Kane, *The Second Billion: Population and Family Planning in China* (Ringwood, Victoria, AU: Penguin Books, 1987), p. 52.

55. Ibid., p.19

56. Quoted in ibid., p. 58.

57. Ibid., p. 70.

58. Ibid., p. 78–79.

59. Ibid., pp. 86–87.

60. John S. Aird, *Slaughter of the Innocents: Coercive Birth Control in China* (Washington, DC: AEI Press, 1990), pp. 93–94.

61. Suzanne Staggenborg, *The Pro-Choice Movement* (New York: Oxford, 1991), p. 17.

62. Quoted in Laurence H. Tribe, *Abortion: the Clash of Absolutes* (New York: Norton, 1990), p. 149.

63. Faye Ginsburg, "Rescuing the Nation," in *Abortion Wars,* ed. Rickie Solinger (Berkeley: University of California Press, 1998), p. 231.

64. Diane B. Paul, *Controlling Human Heredity* (Atlantic Highlands, NJ: Humanities Press, 1995), p. 28.

65. Charles Darwin, *The Descent of Man* (Princeton: Princeton Univ. Press, 1981), p. 403.

66. Henry Herbert Goddard, *The Kallikak Family* (New York: Macmillan, 1913).

67. Paul, *Controlling Human Heredity,* pp. 81–82.

68. *Buck v. Bell,* 274 U.S. 200 (1927).

69. Theodore Roosevelt, "Twisted Eugenics," in *The Works of Theodore Roosevelt* (New York, C. Scribner's Sons, 1926) National Edition, vol. XII, *Literary Essays* p. 201. Originally published in *Outlook,* January 3, 1914.

70. Ibid., p. 44.

— 5 —
Nuclear War: Secret Planning

Of the threats covered in this book, nuclear war is by far the most apocalyptic. The scientists who built the atom bomb recognized its danger. In 1947 the *Bulletin of the Atomic Scientists* published the first Doomsday Clock on its cover to dramatize the countdown toward the apocalypse. It was set at seven minutes to midnight. Two years later, the minute hand crept forward to three minutes to midnight when the Soviet Union exploded its first bomb, and in 1953 the clock went to two minutes to midnight when the United States and the Soviet Union both tested hydrogen bombs. In 1960 it went back to seven minutes before midnight as world public opinion and scientific exchanges gave hope of negotiation, and three years late went to 12 minutes with the signing of the Partial Test Ban Treaty. In 1984 the *Bulletin* set it forward to three minutes to midnight as the arms race accelerated. In 1991 the collapse of the Soviet Union, the end of its domination of eastern Europe, and the signing the Strategic Arms Reduction Treaty relaxed the threat, so it was set at 17 minutes. In 2002 it moved forward to seven minutes as the United States withdrew from the Anti-Ballistic Missile Treaty and terrorists sought to acquire nuclear weapons. In 2007 the clock moved to five minutes to midnight.

Writing in 1957, Henry Kissinger, the future secretary of state, described how an attack on the United States "would produce 15 to 20 million dead and 20 to 25 million injured."[1] Peter Goodwin estimated, "Nine Americans out of every ten could perish in an all-out nuclear attack against the United States."[2] The Royal Swedish Academy of Sciences warned that "...the impact of a nuclear war would be far more devastating to the biosphere than any other threat that is likely to appear in our time."[3] Carl Sagan decried "the threat that nuclear winter portends—a global climatic catastrophe and the deaths of billions of people...."[4]

From its beginnings in the Manhattan Project of World War II, atom bombs have been shrouded in secrecy. Even Harry Truman did not know of the existence of the program until the day after he had been sworn in following Roosevelt's death. Furthermore, the strategy of deploying the weapons was a combination of deliberate planning and simply manufacturing more and more weapons. Although Congress in 1946 established the civilian Atomic Energy Commission to take over the functions of the Manhattan Project from the army, the new agency cooperated closely with the military and maintained strict secrecy.

American foreign policy planning was comprehensive, not just for military action. The global approach was to contain the Communist threat. The planners assumed that cooperation was impossible because the Communists wanted to destroy all democratic countries and impose their own system. But they did understand raw power, hence the United States, Britain, France, and others allies should form military alliances like NATO. The Western countries would not attack the Communists, but neither would they permit aggression as in Korea. The Truman administration developed this secret plan in 1950, calling it NSC 68. The initials stood for the National Security Council, an elite group established by the National Security Act of 1947, the same law that established the Department of Defense and the Central Intelligence Agency.

THE ARMS RACE

The arms race was the unintended consequence. After the Soviet Union exploded its own atom bomb in 1949, the United States built up its arsenal, adding more each year. The invention of the hydrogen bomb in 1952 created a weapon a thousand times as powerful. By 1954 the United States had 2,000 bombs, compared to 150 for the Soviets. The air force did not really have a strategy. If war came, it intended to attack the USSR with all its bombers once, nicknamed the Sunday Punch (to use boxing terminology for a knockout blow). President Eisenhower's secretary of state, John Foster Dulles, labeled the scheme "massive retaliation." If the USSR attacked vital American interests anywhere in the world (for instance Europe or the Philippines), the response would be total. The strategy had the added benefit of saving money in the government budget because bombs were cheaper than a large army and navy. The nickname was "more bang for the buck." To amplify this belligerent stance, Dulles publicly described how it was sometimes a good tactic to push an enemy to the brink of war to intimidate him, a method soon branded "brinkmanship."

The Eisenhower National Security Council tried its own hand at comprehensive planning, in what eventually became NSC 162. Its scope was broader than NSC 68 five years before in that it attempted to assess the impact of high military spending on the civilian economy. While the intention was noble, in the end it proved impossible to integrate the two sides, and the report had to settle for some vague language of compromise. From the first days of the Manhattan Project, it was obvious to the scientists that defense against atom bombs was impossible. The only way to prevent an attack was to threaten to destroy the enemy, eventually labeled mutually assured destruction.

Although the killing of military and civilians was the greatest fear, environmental damage attracted attention. Besides its fearsome blast, the atom and hydrogen bombs produced equally fearsome radiation. The explosion creates tons of radioactive debris that rises into the air with the heat of the mushroom-shaped cloud, then falls to earth downwind, some close by, but with other particles that stay airborne for miles, even thousands of miles. At Hiroshima, the number who died of radiation sickness was equal to the number who died of the blast itself. After the war, radioactive fallout from atomic tests became a problem.

Timeline for Nuclear War

1939	Albert Einstein writes Pres. Roosevelt that an atomic bomb is possible.
1942	Manhattan Project underway to invent and build an atomic bomb.
1945	Atomic bomb tested in New Mexico and dropped on Hiroshima and Nagasaki. Federation of American Scientists organized, originally called Atomic Scientists.
1946	Congress establishes the Atomic Energy Commission.
1949	USSR tests an atomic bomb.
1952	United States tests the first hydrogen bomb at Enewetak Atoll in the Pacific.
1953	Pres. Eisenhower announces Atoms for Peace.
1954	Secretary of State Dulles describes massive retaliation.
1957	International Atomic Energy Agency established by the United Nations.
1960	France tests a nuclear bomb.
1962	Cuban Missile Crisis almost leads to nuclear war.

1963   Partial Test Ban Treaty ends testing above ground.

1964   China tests a nuclear bomb.

1965   Israel estimated to have nuclear weapons.

1968   Non-Proliferation Treaty signed.

1972   ABM Treaty and Interim Agreement signed (SALT I).

1974   Atomic Energy Commission reorganized into ERDA. India tests a nuclear bomb.

1977   ERDA reorganized into the Department of Energy.

1979   United States and USSR negotiate SALT II, but the United States does not ratify because Soviets invade Afghanistan.

1981   Israeli aircraft destroy Iraqi reactor experimenting on nuclear weapons.

1983   Pres. Reagan proposes Strategic Defense Initiative, labeled Star Wars.

1986   Chernobyl nuclear reactor in USSR explodes, spreading radioactivity across Europe.

1989   Berlin Wall torn down, marking the end of the Cold War.

1991   USSR and United States sign START I to limit nuclear warheads.

1993   Russia and United States sign START II to further limit warheads.

1998   Pakistan tests a nuclear bomb.

2002   Russia and United States sign Strategic Offensive Reductions Treaty in Moscow. United States withdraws from ABM Treaty. Pres. Bush announces Ground-Based Missile Defense (Star Wars redux).

2006   North Korea tests a nuclear weapon.

———————————

Once regular tests began in Nevada, radioactivity became a problem. Although it would occasionally drift over the few ranches and tiny cross-road towns on the desert, the scientists dismissed the problem, and local people were ignorant of the danger. Indeed the army deliberately exposed soldiers to the blasts to see how they could conduct military operations in conjunction with atomic weapons. Soldiers would be stationed in fox-holes as close as a mile from ground zero, then after the blast would maneuver through the crater wearing goggles and protective clothing. The Atomic Energy Commission invited schoolchildren from Las Vegas to come to watch. The 1953 tests, code named Upshot-Knothole, produced more radiation than predicted. One explosion dropped six roentgens on the town of St. George, Utah, exceeding the safe dose by at least 50 percent. Ranchers reported that sheep downwind were sick, and AEC veterinarians found some suffered from lesions. A long-range effect came to light when, after a few days, heavy rain dumped two roentgens in Troy, New York.[5]

The Nevada proving ground used for atomic weapons was too small for hydrogen weapons, so the following year the AEC moved to the South

Table 5.1
Nuclear Weapon Stockpiles, 1945–2006

Year	U.S.	Russia	UK	France	China	Total
1945	6					6
1949	235	1				236
1953	1,436	120	1			1,557
1954	2,063	150	5			2,218
1960	20,434	1,605	30	1		22,069
1964	30,751	5,221	310	4	1	36,287
1970	26,119	11,643	280	36	75	38,153
1975	27,052	19,055	350	188	185	46,830
1980	23,764	30,062	350	250	280	54,706
1986	24,401	45,000	300	355	425	70,481
1990	21,004	37,000	300	505	430	59,239
1995	12,144	27,000	300	500	400	40,344
2000	10,577	21,000	185	470	400	32,635
2006	10,104	16,000	200	350	200	26,854

Source: Adapted from Robert S. Norris and Hans M. Kristensen, "Nuclear Notebook," *Bulletin of the Atomic Scientists* 58 (2002):103–104; idem 62 (2006): 64–66.

Pacific for its series code named Castle. The site was the Bikini atoll, and weather conditions seemed perfect. The Bravo explosion was more than twice as powerful as expected, 15 megatons, and radiation reached 25 roentgens. At the last minute, the weather deteriorated and a wind blew to the east. The navy hurriedly evacuated natives off islands as far as two hundred miles distant, but even with decontamination, some of them became sick. Unbeknownst to the Navy, a Japanese trawler, the *Lucky Dragon,* was fishing 82 miles east of Bikini. The fishermen were both ignorant of fallout and afraid the U.S. Navy might detain them, so they did nothing to decontaminate their bodies. When they reached their home port in Japan two weeks later, all 23 in the crew were suffering from radiation sickness and required hospitalization. One died. The Japanese people and government were outraged. To them it seemed a repeat of Hiroshima and Nagasaki.

The dangers of fallout brought pressure to end testing. Scientific reports measured the immediate sickness and the longer-term threats such as from strontium 90 that caused bone tumors and leukemia. Eisenhower investigated ways to cooperate with the Soviets to stop testing. He appointed Harold Stassen to work full time on it and announced that his post was of cabinet status. He was dubbed the secretary of peace. Negotiators met at London for several years, but were unable to reach an agreement. In a further complication, France successfully tested its own nuclear weapon. Eisenhower's other attempt to deal with the atomic danger was called Atoms for Peace, a plan to cooperate with other countries for electric power plants, merchant ships, and medical isotopes. He supported the establishment of the International Atomic Energy Agency (IAEA) under the auspices of the United Nations.

Even though the scientific details of fallout from tests were closely guarded by the military and the AEC, the general risks were well known to the public. Nine thousand scientists had signed a petition to stop testing. The *Bulletin of the Atomic Scientists* frequently ran negative articles. A group calling itself the National Committee for a Sane Nuclear Policy (SANE) organized with the specific goal of ending testing. The answer appeared to be underground testing. By digging tunnels deep beneath the Nevada desert, virtually no radioactive material would be ejected into the atmosphere, hence none would fall out. Even the AEC could see practical benefits. Tests could be scheduled throughout the year without waiting for the annual series. The weather would never be unfavorable. The expenses to dig a tunnel were no more than to build a tower. Large explosions would not have to be conducted in the Pacific.

Eisenhower ended his eight years as president with a farewell address with a surprising conclusion for a man who had served 36 years as an army officer, in which he warned of the dangers of the military industrial complex, terminology he coined in the speech. Ike observed that in the 1950s the United States had created "a permanent armaments industry of vast proportions." For its own profits, the industry encouraged the government to develop weapons. He warned that

In the councils of government, we must guard against the acquisition of unwarranted influence, whether sought or unsought, by the military industrial complex.

The potential for the disastrous rise of misplaced power exists and will persist. Eisenhower feared the "danger that public policy could itself become the captive of a scientific-technological elite."[6]

Herman Kahn, a physicist at the Rand Corporation, shared similar concerns. He complained that he found it hard to interest people in the details of nuclear strategy and tactics, even though many were interested in the politics leading up to a war. His 1960 book, *On Thermonuclear War,* set out the details of the number and size of cities targeted, the number of bombers, and the size of the bombs. He was particularly worried about escalation, by which he meant how one act would lead to moving up to ladder to a stronger one. For example, if the Soviets threatened a conventional attack on Berlin, the Americans might drop a small atom bomb on the USSR, and then the Soviets might drop a bomb on an American city. The retaliation could escalate to total nuclear war. On the other hand, Kahn pointed out that this did not necessarily have to follow. Perhaps the United States would not retaliate after absorbing the blow. Kahn found the outbreak of World War I an important lesson. In 1914 the mobilization plans of the Germans, Russians, and French were so rigid that once the process began, it could not be stopped. No one actually wanted war; they only wanted to bluff. But the consequence was a war greater than ever imagined.[7] In response to criticism that nuclear war was unthinkable, Kahn titled his next book *Thinking about the Unthinkable.*

After John F. Kennedy became president in 1961, conditions became more favorable. To enhance public confidence, he announced that the United States would never strike first with its nuclear weapons. On the other side, Khrushchev gave a speech rejecting general war as a way to victory for socialism. Instead he said the USSR would support wars of liberation in the third world as a means to overcome the capitalist nations.[8] In the prior year, the Soviets had supported the new Castro regime in Cuba and the Communists in the Congo (newly independent from Belgium) and soon stepped up their support of the Viet Cong in South Vietnam. While these little wars were not good, at least they did not threaten an instant Armageddon. As another option, Kennedy ordered the air force to retarget its missiles away from cities, and exclusively against military sites. At Kennedy's request, Congress established the Arms Control and Disarmament Agency. To discourage allies like France from developing their own nuclear weaponry, the administration proposed a multilateral force, whereby the NATO allies would share in nuclear war planning and decisions. Some negotiations with the Soviets continued, now being held in Geneva rather than London.

In October 1962, the strategy of brinkmanship reached its logical culmination in the Cuban Missile Crisis. As the price for Soviet friendship and aid, Fidel Castro agreed to let the USSR station intermediate-range ballistic missiles on his island. They had a range of 1,100 miles, allowing them to target Washington, DC. Khrushchev's rationale appears to have been that he could show support for Castro and intimidate the West. A collateral benefit was to gain prestige at the expense of Communist China, which was becoming a rival. The first phase in Cuba was to deploy antiaircraft installations and fighter planes to protect the missiles, after which the second

Table 5.2
Treaties on Nuclear Weapons

Year	Treaty	Major Parties	Description and Comments
1963	Partial Test Ban	USSR and U.S.	Ban nuclear tests above ground.
1968	Non-proliferation	Over 100 countries	Countries without nuclear weapons promise not to get them. China and France did not sign.
1971	Accidents Measures Agreement	USSR, U.S., U.K.	Prevents nuclear weapons accidents.
1972	Anti-Ballistic Missile Treaty	USSR and U.S.	Outlaws defense against ICBMs. U.S. withdrew in 2002.
1972	Interim Agreement (SALT I)	USSR and U.S.	Limits the number of nuclear warheads.
1974	Vladivostock Agreement	USSR and U.S.	Further limits on warheads; not a treaty.
1979	SALT II	USSR and U.S.	Never ratified. U.S. withdrew. Both side abided with it, however.
1987	Intermediate-Range Nuclear Forces	USSR and U.S.	Limit intermediate-range nuclear forces.
1991	START I	USSR and U.S.; later Ukraine, etc.	Limits nuclear warheads.
1993	START II	Russia and U.S.	Limits ICBMs.
2002	Strategic Offense Reductions	Russia and U.S.	Limits offensive nuclear warheads.

phase was to deploy the missiles themselves. The 10 installations scattered around the island would be capable of firing 80 missiles, each having a warhead of one megaton, that is, a hundred times as much as Hiroshima.[9]

On October 15 aerial photographs showed clear evidence that the mysterious activity observed during the previous weeks was in fact missile facilities under construction. Surveillance of ship traffic, refugee reports, and spy accounts came together with earlier photographs to confirm the enterprise. Kennedy quickly called a meeting of what became known as the Ex Comm: the Executive Committee of the National Security Council. Its membership consisted of the secretary of state, the secretary of defense (McNamara), the director of central intelligence, and the assistant secretary of defense for security affairs. The attorney general, Robert Kennedy, joined because he was the president's brother.

The Ex Comm needed to determine what the American goal was. Was it merely a Soviet promise not to shoot the missiles at the United States? Was it to remove the missiles? Was it to overthrow Castro and return Cuba to a pro-American government? Attention quickly centered on the removal of the missiles.[10] Next the Ex Comm needed to find a means. A military invasion, the only sure cure, would result in thousands of American soldiers killed, and possibly the successful launching of a missile against an American city, with hundreds of thousands of deaths. The air force offered a plan to bomb the missile sites, but it could not guarantee that it would wipe out 100 percent so that none could launch a missile. The navy could blockade the island to prevent new missiles from arriving. Diplomats could negotiate by offering items such as American withdrawal of missiles in Turkey or changes in the status of Berlin. Finally the Ex Comm directed attention to communication with the Soviets. How would it send messages? What would they say? What would be the answers to messages from Khrushchev?

Kennedy and his Ex Comm decided that the American goal was to force the Soviets to remove their missiles. Replacing Castro appeared unattainable, and leaving the missiles was too dangerous. The chief method was to be a naval blockade, which was less risky than an invasion and more powerful than diplomatic talk by itself. Bombing, even a so-called surgical air strike that would only destroy the missile sites, would leave the United States labeled as an aggressor in Communist propaganda for years to come. The committee suggested calling the blockade a quarantine in order to appear less a violation of international law of the seas and to get the Organization of American States to vote in favor of it. The Ex Comm proved useful in advising on communication. For instance, when Khrushchev sent two messages the same day, one emotional and moderate, and the second formal and demanding, the committee suggested answering the moderate one and ignoring the more demanding one.

Escaping Armageddon in the Cuban Missile Crisis strengthened Kennedy's desire for a test ban treaty. It both showed the danger of how close the two counties had come to nuclear destruction and displayed the president's resolve to use the weapons if necessary, thereby dispelling any image of weakness both to the Soviets and to American critics in conservative political circles and in the military. The experience had proved sobering for Khrushchev as well. In July 1963 American diplomats (joined by the British) went to Moscow to negotiate and quickly reached agreement to end testing in the atmosphere. Underground testing was still permitted.

During the entire process, Kennedy was conscious of the need to gain support for the test ban from the public, and especially from the Senate, because it had to ratify the treaty. Earlier he had established the Arms Control and Disarmament Agency by legislation rather than simply by a reorganization plan within the executive branch (for which he had the legal authority), because he wanted to enhance its prestige and publicize its goals. He gave several speeches about banning tests, the most famous one at American University. The White House encouraged the creation of the Citizens Committee for the Nuclear Test Ban Treaty consisting of prominent citizens of both political parties and including two former cabinet members from the Eisenhower administration. This paid off within two months, when the Senate ratified the treaty by a vote of 80–19.

The Test Ban Treaty yielded two successes. First was the end to pollution of the atmosphere by radioactivity. From then on, the United States, the USSR, and Britain completely ended above-ground tests. Two new nuclear powers, France and China, who did not sign the treaty, conducted a total of 60 atmospheric tests in the following years. The other success was the negotiation of further treaties, starting with the Non-Proliferation Treaty five years later. As Kennedy had hoped, this was the first step. On the negative side, the Test Ban Treaty certainly did not end testing or the arms race, which moved underground. From 1963 to 1990 the United States detonated six hundred bombs and the Soviet Union detonated four hundred bombs. This was a much higher rate than before the treaty.[11]

NON-PROLIFERATION

At the height of the Cold War, the United States and the USSR negotiated a series of treaties limiting the number of weapons and their spread to additional countries. The Partial Test Ban Treaty was the first, followed by those on non-proliferation, safeguards against accidents, limits to anti-ballistic missile defenses, limits to the number of launchers, and limits to the total numbers of weapons.

Even as they were developing the atom bomb during World War II, the Manhattan Project scientists anticipated the problems of proliferation. As more countries got weapons, they would be harder to control. The Soviet Union was the first to break the American monopoly, an eventuality impossible to prevent. Although Great Britain was a partner in the Manhattan Project and was supposed to have access to the technology, in fact after the war the Americans did not cooperate, so the British had to develop their own bomb, which they tested in 1952. Because the Americans and British already cooperated so closely on foreign policy, the Soviets did not see this as an increased threat. France independently developed an atom bomb. From the Soviet perspective, this was more troublesome than the British bomb, but still not a direct threat, since it already faced France in Europe as part of NATO. Moreover, France tended to be slightly more favorably disposed and had a large Communist Party.

When the People's Republic of China exploded an atom bomb in 1964, both the Soviets and the Western powers felt menaced. Although the USSR had backed its fellow Communists in taking over the country in 1949, a decade later relations had soured. Marxist ideology was not enough; the geopolitical realities of a country with 700 million people on its border generated fear. When the Chinese had asked for technical help with nuclear energy, the Soviets had refused. Nevertheless, the Chinese had been able to build their own device.

Another country with the technical and financial capability to build an atom bomb at the time was Israel. The tiny nation was endangered on all sides by its Arab neighbors. In 1955 it approached France, playing on their common enemies, especially the Egyptians, who supported the Algerians. France provided a reactor, located near Dimona in the desert, complete with a secret underground plant that produced weapons-grade plutonium.[12] The Israelis never officially admitted their ability to build weapons, especially since the United States threatened to cut off financial aid; yet they did not deny possessing the bombs. They claim that they do not have "operational weapons," but it only takes a few minutes to change that. The Israeli position is that they will not be the first to use nuclear weapons in the regions, but will not be the third.

Several other countries were on the track toward joining the nuclear club, though less far along. In 1956 Canada sold India a civilian reactor that had the potential to produce plutonium. Although impoverished, India could scrape together enough money and had many highly trained scientists. On several occasions it had fought its neighbor Pakistan and in 1962 fought China in the Himalaya Mountains. South Africa at the time had an embattled white government trying to maintain its apartheid policy of racial segregation. Much evidence existed that it was cooperating with Israel. Brazil had the technology and money and was in the process of purchasing six to eight reactors from West Germany, along with a reprocessing plant and a uranium enrichment plant that would encompass a complete fuel cycle.[13] This capability stimulated Argentina to begin its own effort. In the more distant future, potential nuclear states seemed to include Pakistan, Taiwan, and South Korea. The USSR worried about West Germany, even though it did not intend to build a bomb. The Soviets could not put aside their fear and hatred based on World War II.

As the USSR and the United States became more aware of their common interests in a non-proliferation treaty, the U.S. government position was not completely unified. The Defense Department and the Arms Control and Disarmament Agency were the strongest advocates, believing that the risks of proliferation were great. The State Department was less eager as it continued to promote the multilateral force and maintained that, without it, the European allies would see a treaty as disregarding their interests. On the practical side, the Atomic Energy Commission and the Arms Control and Disarmament Agency wanted safeguards by the International Atomic Energy Agency, whereas the State Department favored Euratom, whose members were all from western Europe.[14]

On July 1, 1968, the United States, the USSR, and Great Britain signed the Treaty on Non-Proliferation of Nuclear Weapons, joined by nearly a hundred countries without weapons. The first three pledged not to transfer weapons or to assist others to manufacture them, and the other non-nuclear states pledged not to build or receive any nuclear weapons. Two additional weapons countries (France and China) refused to sign. Several other countries also refused, including ones close to developing their own capabilities: India, Pakistan, Israel, South Africa, Brazil, and Argentina.

SALT, THE ABM TREATY, AND NUCLEAR WINTER

In 1962 the USSR began to build a crude anti-ballistic missile defense around Leningrad, then suspended it, but two years later began construction of an ABM system around Moscow. The United States feared this was the beginning of a nationwide defense system that would neutralize an American attack. One possible response was for the United States to build its own ABM system, but that appeared technically impractical as well as expensive. Another response was to develop MIRVs: multiple independently targeted reentry vehicles. This technique of putting more than one weapon in the same missile, whereby each warhead could hit a different target, made defense much harder. Both nations began MIRVing their missiles, causing a major escalation of the arms race. At the same time, the USSR stepped up its rate of missile manufacturing. Although the Soviets soon abandoned their two ABM attempts, the problems of missile defense and more weapons remained.

In 1966 President Lyndon Johnson directed his ambassador in Moscow to approach the Soviets about negotiations limiting nuclear weapons, and they agreed, although the talks did not begin until June 1968, and then were suspended when the USSR invaded Czechoslovakia two months later. Within the Kremlin, the political situation was in flux. In 1964 Khrushchev was removed from power because of his recklessness in Cuba, failures in agriculture, and the threat of China. Over the next few years Leonid Brezhnev emerged as the leader, and the policy of confrontation gave way to one of cooperation, labeled detente.

The next president, Richard Nixon, and his National Security Council director, Henry Kissinger, revived the strategic arms limitation talks (called SALT), believing them to be a good way to constrain the Soviets, who were rapidly increasing their arsenal. In 1967 when Johnson was trying to get the talks going, they had 500 missiles, compared to 1,630 for the Americans, while two years later when Nixon actually began the talks, the Soviets had 1,900 missiles, and when they concluded three years later, they had 2,348.[15] The talks took place in Helsinki and Vienna. Their content was kept secret from the public to prevent divulging the details of the technology. The parties reached agreement on both limiting weapons and prohibiting an ABM system. When Nixon journeyed to Moscow the following spring for a summit meeting, he and Brezhnev signed the two treaties.

During the SALT, both sides contemplated how the nuclear issue was linked to other foreign policy objectives. The USSR was officially supporting its fellow Communist regime in North Vietnam in its war against South Vietnam and the United States. On the other side, the United States was cultivating better relations with the People's Republic of China, which the Soviet Union was trying to isolate. In February 1972 Nixon stunned the world by paying an official visit to Beijing. The Prague Spring of 1968, in which the Czech government succeeded for a few months in acting independently of Soviet domination, gave great hope to the United States and its European allies that Eastern Europe would soon shake off its Soviet masters. In fact, the linkage was surprisingly small. The USSR did not set up conditions based on Vietnam or China. The United States did suspend the talks temporarily in August 1968 when the Warsaw Pact troops invaded Czechoslovakia to crush the democratic government, but renewed them a year later.

The success of SALT encouraged more negotiations, this time aimed at reducing the total number of weapons, not just freezing them at existing levels. A criticism of the treaty was that it merely formalized the status quo. The 1972 treaty, which was not permanent, also left open issues of verification and new sorts of missiles. Nixon, who enjoyed a reputation for skill in foreign policy, resigned in August 1974 as a consequence of the Watergate scandal, but his successor, Jerry Ford, kept most of his policies, and his advisor, Henry Kissinger, by then was also appointed as secretary of state. The new president met Brezhnev at a summit meeting in Vladivostok, where the two signed an executive agreement (not a treaty) setting a limit of 2,400 for missiles and bombers, and a limit of 1,320 for those equipped with MIRVs. This was intended to eventually be the basis of a formal treaty.

Talks did not go well, however. Verification for SALT provisions proved difficult. The Soviets appeared to be cheating by increasing the payload of the SS-19 missile and by putting nuclear bombs in a small bomber aircraft. They began testing their own MIRVs. On the American side, the U.S. Air Force introduced a small cruise missile that did not fit the existing categories. Linkages, which the two countries had tended to ignore for SALT I, now came to the forefront. The USSR wanted the economic trade advantages, but Congress was reluctant to vote for this. A number of senators wanted to make the economic concessions depend on the Soviets permitting freer emigration of Jews from the USSR. In the presidential primary elections, Ford encountered stiff opposition from Governor Ronald Reagan, who accused him of giving away too much to the Communists. Certainly a further problem was that once the first agreement was signed, secrecy was a less effective source of power. During the SALT negotiations, most people were not aware that the talks were taking place, but afterwards, their success brought more attention to the issue. Of course, the details of the technology were still secret, but knowledge of types of weapons and their capabilities became more widespread. Moreover, in the aftermath of the Watergate scandal, the public became skeptical about presidential actions. People were not giving him the benefit of the doubt or the freedom to explore options.

Within weeks of becoming president, Jimmy Carter tried to revive the negotiations, called SALT II, proposing deep cuts in the number of missiles. He also approached senators and other critics who had opposed the Vladivostok formula to gain their support. The Soviets did not want to make the deep cuts Carter proposed. The new secretary of state, Cyrus Vance, negotiated directly with the Soviet ambassador in Washington, Anatoly Dobrynin, to reach a three-part compromise. A formal treaty lasting until 1985 would codify the Vladivostok agreement and incorporate a few reductions. A three-year protocol would address controversial issues, and a joint statement of principles would set the agenda for future negotiations. Not until June 1979 did Carter and Brezhnev sign the treaty-protocol-principles at a summit meeting in Vienna.

Both the Nixon-Ford and the Carter administrations tried to link SALT to nonmilitary issues. Kissinger made it the centerpiece for detente. He wanted to use it to discourage Soviet aid to North Vietnam, to divide the USSR from China, to restrain it in supplying arms to Communist guerrillas in Angola and the Horn of Africa, and to loosen restrictions for Jews who wanted to emigrate. Kissinger believed in 1971 that the Soviets would be more cooperative because they wanted to buy American grain to make up for their bad harvests. Carter wanted to use SALT to encourage Soviet restraint in aiding guerrillas in Africa and to permit emigration of Jews. In December 1979, after Carter and Brezhnev had signed their agreement in Vienna, but before the Senate had ratified it, the USSR invaded Afghanistan with regular army troops. Carter attempted to use the treaty as a means of forcing them to withdraw, but this was unsuccessful. In January, he asked the Senate to defer consideration of the treaty as long as the Soviets occupied Afghanistan. On the other hand, the USSR did not pay much attention to linking SALT and other foreign issues. It had not curtailed negotiations in the early 1970s when the United States increased bombing of its Communist ally, North Vietnam, or when Kissinger and later Nixon visited China.

Jimmy Carter himself was well aware of the dangers of nuclear war, having spent most of his eight-year naval career assigned to the new nuclear submarine program. He personally detested nuclear bombs and strongly supported SALT. Nevertheless, his administration was unable to accomplish much and suffered from several blunders on weapons decisions. Early in his term of office, the Pentagon made plans to introduce a new artillery weapon into Europe, known officially as the enhanced radiation warhead and unofficially as the neutron bomb. The nuclear artillery shell was designed specifically to kill enemy troops by radiation with less physical damage from the explosion. The Soviets called it the capitalist bomb because it did not damage property. All the warheads would be deployed in West Germany, which caused opposition from the people and leaders. After twisting the arm of the German Chancellor to give his grudging acceptance, Carter abruptly canceled the project at the last minute.

Ronald Reagan won election in 1980 after a campaign to be tough to the Communists. Reagan called the USSR an Evil Empire and declared that its leaders "reserve unto themselves the right to commit any crime, to lie, to cheat."[16] This fit the president's old-fashioned religious faith, which was often literal. On several occasions he said that he believed that the world could end during his lifetime and declared during the 1984 presidential debate that it was possible that Armageddon could happen tomorrow. Officials like Secretary of State Alexander Haig and Secretary of Defense Caspar Weinberger talked openly of the how to win a nuclear war. Haig spoke about the possibility of firing a "nuclear warning shot" inside Europe.[17] The new secretary of the navy said that the United States should not be bound by SALT I or II. In the Pentagon, Weinberger's assistant secretary for international security policy, Richard Perle, was a vociferous critic of nuclear weapons control. Meanwhile, the Pentagon began buying more and more weapons. During the four years of Reagan's first term, military spending rose 40 percent. While the tenor of the Reagan administration was hostile toward negotiations, his foreign policy team was not well coordinated. Haig and Weinberger frequently clashed. Haig himself, a former career soldier as well as a protégé of Henry Kissinger in the Nixon-Ford White House, felt frustrated by the president's unwillingness to give clear directions and by the ineffectiveness of the National Security Council staff. Haig tried to emulate his former mentor by taking the initiative in policy declarations and attempting to stifle the opposition, but was not very successful. Across the Potomac River in the Pentagon, the chairman of the Joint Chiefs of Staff, Air Force General David Jones, supported arms control, in opposition to his boss. The Joint Chiefs preferred the stability of SALT rather than a wide-open arms race. The new administration changed the name to START for Strategic Arms Reduction Talks.

Simultaneously with Reagan's ascension to the presidency, members of the public who wanted an end to the nuclear arms race mounted a robust movement for a nuclear freeze. The American Friends Service Committee asked Randall Frostberg, then a graduate student in political science at MIT, to draft a statement in 1979. Frostberg penned a "Call to Halt the Arms Race" that called for a "mutual freeze on the testing, production and deployment of nuclear weapons." The rationale was that a freeze was simple, practical, and understandable, in contrast to the more esoteric negotiations of SALT that depended on expertise and secrecy. The next fall, Reagan's campaign and election galvanized the movement. Over the following two years citizen protests took the form of mass demonstrations, culminating when 750,000 people rallied in New York City in June 1982. Smaller groups participated in other demonstrations, in teach-ins, in picketing, and in lobbying. Press attention increased as the number of magazine and news articles went from a few dozen a year to hundreds.[18] Roman Catholic archbishops and mainline Protestant churches endorsed the freeze, as did the League of Women Voters. When ABC broadcast a television drama titled *The Day After* about life in Lawrence, Kansas, in the aftermath of war, a hundred million viewers watched.[19] On Capitol Hill Democratic Senator Edward Kennedy joined Republican Mark Hatfield and 18 co-sponsors to introduce a resolution of support, while in the House, Representatives John Bingham and John Markey recruited more than a hundred co-sponsors for their resolution. In Europe advocates of the freeze lobbied their own governments and held large rallies.

Further criticism came after the TTAPS report predicted that a full-scale war would create a nuclear winter. Nuclear winter is the phenomenon in the aftermath of a nuclear war whereby the temperature of the earth cools as much as 15–20 degrees F, causing crop failure. The proximate cause is the smoke from fires of cities and oil wells. It involves "darkening, cooling, enhanced radioactivity, toxic pollution, and ozone depletion." The amount of sunlight penetrating the haze would only be 1 percent of normal for the summer, the equivalent of December, hence the name. The terminology came from an article by Turco, Toon, Ackerman, Pollack, and Sagan published in 1983, known as the TTAPS report after the authors' initials.[20] The scientific predictions were based on the study of major volcanic eruptions, but this would be of much greater severity. The artificial winter would lead to the extinction of human beings and the plants and animals they use for food and fiber. At the same time, excessive ultraviolet light would irradiate the planet. Surviving the initial exchange would merely postpone death.[21] The danger was truly apocalyptic.

Sagan was a skilled publicist, and his winter image attracted a lot of attention. While previously many had predicted an apocalypse of 30 million or more killed on each side and further deaths due to long-lasting radiation, TTAPS added the dangers of smoke and soot from fires. Improved computer models of the atmosphere made better predictions possible. Several other scientists replicated the TTAPS study. But in spite of adding a new level of alarm, the warning of nuclear winter did little to change American foreign policy. The consequences of winter were not that much worse than the consequences of 60 million people killed by bombs and radiation.

Sagan's practical recommendation was that the Americans and Soviets both reduce their arsenals to two hundred weapons each. This would be enough to kill millions, hence deter, but not enough to cause the smoke and soot that would reduce sunlight so dramatically.

STAR WARS

Reagan did not pay much attention to the nuclear winter alarmists, nor could he satisfy the advocates of the freeze, but neither could he satisfy his own side. In the 1980 election campaign he had promised a strong defense and criticized the Carter program. At first his administration floundered around for a military policy that would appeal to conservatives. Then on March 23, 1983, he made his dramatic announcement of the Strategic Defense Initiative, immediately dubbed Star Wars after the popular film series of the time. The president sincerely believed that defense was desirable, saying, "The defense policy of the United States is based on a simple premise: The United States does not start fights. We will never be an aggressor." Reagan continued:

Let me share with you a vision of the future which offers hope. It is that we embark on a program to counter the awesome Soviet missile threat with measures that are defensive.... What if free people could live secure in the knowledge that their security did not rest upon the threat of instant U.S. retaliation to deter a Soviet attack, that we could intercept and destroy strategic ballistic missiles before they reached our own soil or that of our allies?[22]

The president proposed a shield over the entire continent that ICBMs could not penetrate, comparing it to a roof on a family's house that protects it from rain. He asked the American people, "Wouldn't it be better to save lives than to avenge them?"[23] The techniques were to be gigantic laser beams projected hundreds of miles above the earth like searchlights, hydrogen particle beams, nuclear weapons exploding in space to disable Soviet ICBMs in mid-flight, and radar on satellites in orbit capable of detecting every missile from the instant it was launched. Satellites pre-positioned in space would shoot lasers and particles; other satellites would be giant mirrors to reflect and aim beams shot from earth to hit Soviet missiles. It amounted to hitting a bullet in flight with another bullet. As Reagan himself admitted, the technology had not been invented yet. Critics quickly labeled it as fantasy. They pointed to a 1940 movie, *Murder in the Air,* in which Reagan had acted, which featured a ray gun that could shoot down distant enemy aircraft.

The speech had been put together by amateurs on the White House staff who had limited knowledge of weapon technology or diplomacy. Neither

Secretary of State George Schultz nor Secretary of Defense Weinberger knew of it more than a day or two in advance, and they were then merely informed, not asked to advise. Apparently the man behind it was the atomic scientist Edward Teller, who was urged on by two personal friends of Reagan: Joseph Coors, the brewery owner, and Karl R. Bendetsen, a businessman.[24] Neither held a government post. In the Pentagon, the Joint Chiefs of Staff had discussed the idea only in a general sort of way as a possibility to be investigated. In the White House, the idea had the backing of Ed Meese, the president's counselor, and William Clark, the new National Security Council director.[25] Neither had a background in international relations nor defense. Both were lawyers who had been with Governor Reagan in California.

The reaction by the experts was negative. Scientists pointed out that the proposed giant laser beams had not been invented, that it was nearly impossible to shoot down an intercontinental missile in flight, and that radar was incapable of tracking every enemy launch. The most serious flaw was that the defense would have to be 100 percent effective; to let even one missile through could mean the death of a million people. An air force general not under the thumb of the secretary of defense told the Senate Armed Services committee that 50 percent of the missiles would still get through. This would amount to 2,500 megatons. Diplomats pointed out that the Strategic Defense Initiative violated the 1972 ABM Treaty. Analysts warned that for it actually to be developed, if possible, would take trillions of dollars over 20 to 30 years. Military officers and civilian bureaucrats considered the idea so outlandish that the project had trouble recruiting staff. Allies in Europe and the Pacific worried that a Star Wars battle would destroy them, and that if the United States actually were to be protected, it might abandon them.

About the only people to take it seriously were the Soviets, who immediately undertook a propaganda campaign in opposition. This backfired, because the Reagan loyalists then argued that if the Soviets were so upset, it must be a good idea. In the propaganda contest, the USSR had the advantage that the SDI appeared to the Europeans as a reckless escalation of the arms race. In the military contest, the USSR had the advantage that it could counter the Americans simply by building more warheads. Indeed, it did not even have to build real warheads; dummy warheads and aluminum chaff would confound the U.S. Air Force defense. In spite of these objections, the SDI proposals enjoyed great popularity with conservative members of the Republican Party throughout the country.

From the American side, the Reagan administration's internal confusion, combined with its hostility toward the USSR, blocked serious negotiations on nuclear arms control. From the Soviet side, the problem was a five-year leadership crisis. Leonid Brezhnev's health deteriorated, and he died in 1982. The Communist leadership replaced him with another sick old man, who lived only 16 months, and then another sick old man, who lived only 13 months. At this point the leaders realized they needed a young and vigorous leader, or else the entire country would wither away. They chose Mikhail Gorbachev, who embarked on a program of openness and reform. The new general secretary wanted to improve the economy and reduce military expenses.

With a strong leader, the Soviets were once again interested in negotiating on nuclear arms control. In November 1985 Gorbachev met Reagan in Geneva. In five hours of private conversation, Gorbachev became convinced that Reagan sincerely wanted to avoid thermonuclear war. The president seemed eager to negotiate. Although Gorbachev tried hard to persuade Reagan to abandon the SDI, he could not budge him. To follow up, in a January speech Gorbachev proposed to abolish all nuclear weapons within 15 years. Next he proposed to dismantle all Soviet and American missiles in Europe. He even said the British and the French could keep their missiles, which had been a sticking point in previous negotiations. The next month Gorbachev addressed the Communist Party Congress at its five-year meeting to say that the USSR should push aside its seven-decade-old dogma of class struggle to recognize that its security depended on cooperation with the United States and other nations. Its best defense against American missiles was to avoid a war.

In April 1986, the Chernobyl nuclear reactor exploded, spreading radioactivity across Europe, thus dramatizing the dangers of the atom and the weakness of Soviet technology. This disaster was due to a series of engineering blunders. During a routine shutdown, some technicians tried an unapproved experiment, which quickly went wrong. The core heated up uncontrollably, and the test team had turned off the emergency cooling system. A small part of the core "went critical," exploding like a small bomb. Four seconds later, another explosion, perhaps nuclear or perhaps steam, blasted the thousand-ton roof off the plant. Because the government was afraid to admit to the accident, the first news came when the Swedes detected high levels of radioactivity blown by the wind. Within a week the cloud of fallout had circled the globe. Eventually 30 people died, 238 were sickened, and thousands were evacuated. Even though Chernobyl was a civilian plant, it cast doubt on all aspects of Soviet nuclear engineering. Once the secret was exposed, popular pressure grew against the old ways and in favor of Gorbachev and his reforms.

Gorbachev met Reagan again in October in Reykjavik, Iceland, where in an extraordinary weekend session, they both threw on the table dramatic proposals for disarmament, most of them poorly thought out (at least on the American side). In a free-wheeling discussion on the first day, Reagan and Gorbachev tentatively agreed to ban all ballistic missiles. Reagan was thinking spontaneously in disregard for the careful preparation of his NSC, State Department, and Defense Department. In the evening, two teams of diplomats began negotiations that lasted until 6:30 A.M. the next morning. They seemed to be making more progress on strategic and intermediate weapons than in the previous 10 years. The problem emerged during the second morning, when Gorbachev told Reagan that the package had to include abandoning SDI. He demanded that the United States formally agree to abide with the 1972 ABM Treaty for a period of 10 years.[26] Although the U.S. Senate had never voted to ratify the treaty, for the previous 14 years the country had acted as if it had. The United States was adhering to an unratified treaty, and the Soviet Union wanted it to continue to do so. Reagan would not accept this, so the summit broke up immediately in an atmosphere of anger and ill will.

The failure at Reykjavik was just the beginning of a string of bad luck for Reagan. Three weeks later in the mid-term congressional elections, the Republicans lost their majority in the Senate, which they had held since the 1980 election. At the end of November, the press published the first reports of what became known as the Iran Contra Scandal. The president's former National Security Council director, Bud McFarland, had secretly traveled to Iran to meet with the leaders of the revolution, who had held the American diplomats hostage for over a year. He was arranging to sell the Iranian government weapons and missiles in return for its help in securing the release of American hostages captured in Lebanon. The proceeds from the sale were to help the Contra rebels in Nicaragua, which the Reagan administration wanted to support in defiance of a congressional ban. One of his staff, Oliver North, had illegally solicited additional money for the Contra guerillas, and worse, had skimmed money for his personal use. The NSC director, Admiral John Poindexter, who was also involved, was forced to resign.

The Americans and the Soviets were negotiating separately to reach a disarmament agreement on intermediate-range nuclear missile forces (INF). Once announced, it did not take long for American, British, and German critics to poke holes in it. In addition the Soviets modified their proposal by demanding that certain German missiles be removed, contrary to their earlier stance. Meanwhile, diplomats in Geneva continued to negotiate. By September the issues were important enough to demand attention at a higher level. The Soviet foreign minister, Eduard Shevardnadze, flew to Washington, and in October Secretary of State George Schultz traveled to Moscow to iron out details so the INF Treaty could be signed at a Washington summit. In his meetings with Schultz, Gorbachev reverted to his earlier hard line that the United States would have to give up the SDI. Shultz returned to Washington with an apparent failure. Later evidence showed that, at a secret meeting of the Communist Central Committee a few days before, Gorbachev had been hit by an attack from both the right and left. Conservative Communists accused him of moving too fast on his reforms, and the reformers, headed by Boris Yeltsin, accused him of moving too slowly. Yet within a few days, Shevardnadze had suddenly flown to Washington to backtrack and arrange a summit to start on December 7. Gorbachev's demands to terminate the SDI had disappeared.

Bargaining on the final details of the INF Treaty went into high gear in Geneva. Last-minute complications required Shultz and Shevardnadze to make a quick trip to Geneva on November 23–24 to reach agreement. Even with this, issues remained unresolved about inspection, so the U.S. and Soviet negotiators bargained until December 7 and then continued in the air as both teams flew across the Atlantic in a U.S. Air Force cargo plane.[27] The following afternoon, in the East Room of the White House, the president and the general secretary signed the treaty. The document provided that the USSR would destroy 1,846 missiles and the United States would destroy 846 missiles. Although Gorbachev did not manage to outlaw the SDI, the two leaders issued a communiqué on the final day saying that they would abide by the ABM Treaty. Furthermore Gorbachev was aware that a few weeks earlier Congress had passed a budget that cut the SDI appropriation by a third and forbade tests in space that violated the ABM Treaty. By withdrawing his rigid opposition to SDI, Gorbachev got the treaty, the communiqué, and a great public relations opportunity. During his time in Washington, he appeared on national television broadcast to both countries, met with many private citizens at his embassy, and shook hands with dozens of ordinary people on the sidewalk when he stopped his limousine at Connecticut Avenue and L Street. Public opinion polls showed 56 percent of Americans approved of Gorbachev, higher than Reagan's 49 percent level at the time. The press called it "Gorby Fever."[28]

Although the two leaders had ended their Washington summit with the pious declaration that they wanted to reach agreement on START, by their next meeting that proved premature. When they met the following May in Moscow, they were able to sign the INF Treaty, which was then ratified by the Senate, unlike the ABM and the SALT II treaties. After this high point, the momentum for nuclear arms control dissipated. In the United States, Reagan was a lame duck in the last year of his term, and the presidential election monopolized public attention.

Back in the USSR Gorbachev had other problems. The openness he proclaimed in 1985 was backfiring as more and more citizens recognized the poor condition of their economy. Plus the reforms had not delivered improvements as quickly as hoped. The military budget was a burden. In July at a conference of the Warsaw Pact, Gorbachev proposed a unilateral reduction of 70,000 men stationed in East Germany, Czechoslovakia, and Hungary, but this was not accepted. At home he proposed cutting the military, but his generals did not accept this. Yet the reluctance of the military was not decisive, as their wishes were overtaken by events. As the year 1989 progressed, Gorbachev's policies of openness and restructuring took on a life of their own.

That spring the USSR held competitive elections for the first time since 1917. Many old-fashioned Communist Party members suffered defeat. People in the Baltic Republics and the Ukraine marched in nationalistic demonstrations, which the police and army were unable to put down. In the satellites of Poland, Hungary, East Germany, and Czechoslovakia, popular demonstrations demanded more freedom. Unlike in 1953, 1956, and 1968, the Red Army did nothing to support the Communist regimes. Poland had legalized the Solidarity trade union and had promised free elections. Hungary dismantled its fence along the border with Austria, permitting free emigration. Worst still, it permitted East Germans to use this route to flee to Austria, thence to West Germany.

George H. W. Bush, who had been inaugurated in January, came to the presidency with greater foreign affairs experience than any president except Jefferson, the Adamses, or Eisenhower. He had been an ambassador to China and the United Nations, the director of the CIA, and vice president for eight years, having visited a total of 70 foreign countries. In spite of this background, he was unable to get his own foreign policy in place promptly and was largely passive as the great events of 1989 occurred. In May his secretary of state, James Baker, met Gorbachev in Moscow, where Gorbachev surprised him by announcing a unilateral cut of five hundred tactical nuclear warheads from the Warsaw Pact countries. Bush tried to regain the initiative by proposing that NATO reduce its troops by 10 percent in a display of competitive disarmament.

In August in Poland the Communists voluntarily agreed to give up power, the first such occurrence in Eastern Europe since the end of World War II. Within days the Hungarian and Czech leaders had promoted reforms. On

November 9 the East German government bowed to the inevitable and opened its gates through the Berlin Wall to permit free emigration. In the delirium of excitement, both East and West Berliners gathered to celebrate and tear down the wall. In Czechoslovakia popular demonstrations in Prague led to the downfall of their Communist government. On Christmas Day the Rumanians overthrew their dictator in the only violent revolution in the six Warsaw Pact countries. The speed of the change had outrun Gorbachev's timetable. He was anxious to meet with Bush. Finding a site proved difficult, but in December the two leaders met aboard naval ships in Malta in the Mediterranean Sea, a summit spoiled by a terrible storm. The subject of nuclear weapons hardly came up. Significantly, Gorbachev declared, "We don't consider you an enemy anymore. Things have changed."[29]

Things had changed domestically for Gorbachev as well. His policies of openness and restructuring were outrunning him. The economy declined. The accident at the Chernobyl reactor had exposed the callousness, manipulation, and incompetence of the government. Once the tragedy became known, Gorbachev played the demagogue in his criticism of the bureaucracy. His economic freedom backfired. His reforms and rhetoric had raised people's hopes, but they remained unfulfilled. As part of his scheme to replace old liners on the Politburo, he had removed the only members representing two Soviet republics: Ukraine and Kazakhstan. Two other republics, Azerbaijan and Armenia, were near a state of war with each other over the disputed territory of Nagorno-Karabakh. Unwittingly he had created a rival when he appointed Boris Yeltsin to head the Moscow Communist Party. Yeltsin had a flair for publicity, amplifying his superior's criticisms of the bureaucracy and the party. To show his independence, he dramatically resigned from the Communist Party.

In March 1991 Gorbachev engineered voter approval of a referendum creating popularly elected presidencies for the various republics. In a vigorous campaign against an old-style Communist, Yeltsin won with 60 percent and took office as president of the Russian Republic on July 10. Only six weeks later, hard-line Communists attempted a coup de état against Gorbachev. The plotters worried that the Soviet Union was about to split into separate republics. Yeltsin courageously fought off the plotters. Barricaded in the Russian Parliament building, he defiantly called a general strike. The next day huge crowds demonstrated in Leningrad and Moscow, where they built barricades at the Parliament building. The coup collapsed, and Gorbachev was saved from defeat. Although Gorbachev was back in power in the Kremlin, his popularity was ebbing. Yeltsin and other reformers demanded a new political arrangement resulting in the dissolution of the

Union of Soviet Socialist Republics. Henceforth each republic would be more autonomous, and indeed they ended up virtually independent. The successor to the former union was a loose confederation called the Commonwealth of Independent States, a name deliberately chosen to avoid the term "union." With no USSR, Gorbachev was out of a job.

As the USSR had begun to split apart and its military threat weakened, President Bush ordered a stand-down of the American nuclear forces. He directed the air force bombers, which had been on alert for 30 years, to curtail their patrols and soon had them unload their bombs. Bush likewise ended the alert status for the strategic missiles scheduled to be eliminated under the provisions of START, which amounted to 450 silo-based Minuteman rockets and the missiles on 10 submarines. Within a week, Gorbachev deactivated five hundred rockets on land and in six submarines, restricted his strategic bombers, and confined his railway missiles to their bases. Both sides withdrew thousands of tactical nuclear weapons in Europe.[30] The United States and the USSR signed START on July 31, 1991, an example of unfortunate timing, since the coup occurred within three weeks, followed by the dissolution of the Soviet Union at the end of December. The treaty was reorganized in the Lisbon Protocol so that Belarus, Kazakhstan, Russia, and Ukraine became the official successors to the USSR. Belarus, Kazakhstan, and Ukraine agreed to give up the nuclear weapons they possessed as part of the old Soviet Union.

The next task became what to do with the missiles no longer deployed. Bush and Gorbachev recognized that they should be destroyed, not merely put in storage. The United States offered to purchase five hundred tons of highly enriched uranium salvaged from the weapons, an arrangement that offered the Soviets an economic incentive. This made a smooth transition to the post-Soviet situation, when an agreement among Russia, Ukraine, Belarus, and Kazakhstan put the old Soviet nuclear arsenal under Russian control. These were the only four republics in which nuclear rockets were deployed. The Russians re-processed the uranium at facilities under the monitoring of Americans. Later presidents Clinton and Yeltsin formalized the arrangement at their summit in 1994. The uranium was shipped to the United States to become fuel in nuclear power reactors. Other fuel was sent to a storage site in the Ural Mountains, where it was to be monitored by the International Atomic Energy Agency staff as well as Americans.[31]

One of the more sensational incidents was named Operation Sapphire. The Central Intelligence Agency learned that six hundred kilograms of uranium were sitting poorly guarded in fabrication plant at Ust Kamenogorsk in Kazakhstan. Iranian scientists had recently visited the facility, perhaps to obtain the uranium for weapons. Immediately the Americans arranged to purchase the material, and within hours of President Clinton's approval, a team of 31 specialists was heading to Kazakhstan. The uranium was flown directly to the United States and sent to Oak Ridge, where it was diluted to a lower enrichment level. The cost of the fuel was estimated to be $10–20 million.[32]

START II was signed by presidents Bush and Yeltsin in January 1993, but was not ratified by the Russian parliament until 2000. When implemented, the treaty was to eliminate heavy ICBMS, including MIRVs. Only ICBMs carrying a single warhead would be allowed. It would also reduce the total number of strategic nuclear weapons deployed by both countries by two-thirds. By the end of the first phase, each side was supposed to reduce its total deployed strategic nuclear warheads to 3,800 to 4,250.

In spite of the disarmament in the 1990s, the two sides still can shoot 5,000 nuclear weapons at each other in half an hour. The United States can launch land-based rockets with 2,000 warheads and submarine-based rockets with 1,000 warheads. The Russians can launch land-based rockets with 3,500 warheads and submarine-based rockets with 300–400 warheads.[33] The warning time for reaction is only 10 or 15 minutes. In both countries the command supposedly can only be given by the president, who is accompanied at all times by a suitcase with the secret codes. The assumption is that the command channels will function normally, and that the president will be ready to get the word and make his decision. The American command system will probably operate satisfactorily, yet while it pays lip service to the constant responsibility being in the hands of the president, in fact a number of military officers have authority to launch an attack, because of the likelihood that in a real war communications may not function fully.

The decrepit state of the Russian Strategic Rocket Forces presents many risks. Many of their radars do not operate. The electronics on the rockets have malfunctioned. The navy can support only two submarines at sea at any given time. Because the rocket bases have not paid their electricity bills, local utilities have cut off their power on several occasions. Housing for the military is poor, and pay is often late. Crews receive less training than before. The railway missiles have been confined to their bases since 1991, and only two regiments of the truck-based missiles can be deployed at a time.[34] In 2001, the Russian missile submarine *Kursk* mysteriously imploded in the Barents Sea. The vulnerable state of their missiles may tempt the Russian generals to strike first, since all of them may be destroyed in the first wave.

The Strategic Offensive Reductions Treaty signed in Moscow in 2002 reduces the number of nuclear weapons. Unfortunately it does not have a method for verification, and it ignores nonoperational warheads. Due to secrecy, estimating the number of weapons is elusive. As of 2006 the United States has 5,735 active warheads. This is out of a total of 10,000 warheads. The remaining ones are inactive, which may mean they are disassembled or may mean that they are merely in storage. Russia has 5,830 active warheads, out of a total of 16,000. China has 200 active warheads. India, which tested its first bomb in 1974, probably has 50–60 active warheads, and Pakistan, which tested its first bomb in 1998, probably has 40–50 active warheads. The estimate for Israel is 60–80 active warheads.[35]

While the nuclear threat from the Soviet Union, and now Russia, has diminished greatly, new problems have emerged. The suicide attacks on the World Trade Center and the Pentagon on September 11, 2001, caused new worries about terrorists obtaining nuclear weapons. The most direct method would be to obtain the actual weapons. The source most often mentioned is renegade soldiers of the Soviet military, but little hard evidence exists that nuclear weapons are missing. The next most likely source is to build bombs from uranium or plutonium acquired from the Soviet military or the Atomic Ministry. While the United States was able to purchase the six hundred kilograms of uranium in Kazakhstan in 1994, many other sources exist. Although they no longer have weapons, Ukraine and Belarus have stockpiles of weapons-grade uranium and plutonium. On nearly a dozen occasions in the early 1990s, police in Russia, Germany, and Czechoslovakia arrested people attempting to sell two or three kilograms of plutonium or uranium. The material most attractive to smugglers is enriched uranium because it is easier to handle and to build a bomb from this than from plutonium.

Besides sources in the former Soviet Union, many worry about the nuclear weapons belonging to Pakistan in the eventuality of a revolution that would install a government favorable to terrorists. Pakistan successfully exploded a nuclear bomb in 1998, after a research program that began as soon as India tested one 24 years earlier. The leading scientist was Abdul Qadeer Khan. Evidence shows that early in the program, he got technical assistance from China. Evidence also shows that Khan collaborated with North Korea, Libya, and Iran.

Although any atom bomb exploded by terrorists would be a catastrophe, a single explosion would not be truly apocalyptic. The world has obviously survived the bombs dropped on Hiroshima and Nagasaki. Using the ominous calculus of the TTAPS report, the number of bombs to cause nuclear winter would have to be in the range of two hundred, which is beyond the capacity of terrorists or rogue states such as Iran or North Korea.

In recent years Iran has begun to develop nuclear weapons. The country signed the Non-Proliferation Treaty in 1968, but that still allows it to develop peaceful, civilian uses. Its leaders routinely claim their experiments are entirely peaceful. However, it is hard to justify their need for nuclear energy when they export 2½ million barrels of oil a day, ranking fourth in the world. A more logical reason is to build a nuclear bomb to use against Israel. While generally paying lip service to the innocence of their goals, a few remarks have slipped out. Only a few years ago, former president Akbar Hashemi Rafsanjani said in a radio broadcast:

If one day, the Islamic world is also equipped with weapons like those that Israel possesses now, then the imperialists' strategy will reach a standstill because the use of even one nuclear bomb inside Israel will destroy everything.[36]

Five years later, evidence existed that Iran was trying to develop a bomb as fast as possible. The International Atomic Energy Agency found highly enriched uranium (well over the 3.5 percent level). This is more than enough enrichment for generating electricity and typical of early stages of weapon manufacturing. Building the bomb is complex and will take 5 to 10 years more. Unlike the U.S.–Soviet confrontation of massive retaliation during the Cold War, Iran seemed to confine its target only to one country, but one so small that it would be decimated. Nevertheless almost certainly Israel could launch a counterstrike that would decimate Iran.

Like Iran, North Korea signed the Non-Proliferation Treaty. Starting in 1984 it built two small research reactors at Yongbyon, which were subject to inspection by the International Atomic Energy Agency. In 1993 it announced it would withdraw from the treaty and denied the IAEA the right to inspect. The United States believed that North Korea was producing plutonium, enough for 10 bombs. It negotiated an "agreed framework" to buy them off with imports of fuel oil and the promise of a light water nuclear reactor. This type is much safer because it does not generate any plutonium. Over the next eight years North Korea did not cooperate fully, so the United States did not follow though on the light water reactors, although it did deliver the oil. North Korea switched from plutonium to producing enriched uranium, suitable for building a nuclear bomb with a different technology. The Bush administration considered this to violate the spirit if not the letter of the agreed framework, and hence terminated it.

North Korean actions and communications were inconsistent and hostile. The dictator, Kim Jong Il, has run the country as a tyrant since inheriting the job from his father in 1994. The economy is nearly bankrupt, and people are literally starving. When foreign diplomats visited the capital of

Pyongyang, Kim sometimes told them he had nuclear weapons and other times told them he did not. On several occasions, North Korea has fired missiles toward Japan. Its missile technology offered an opportunity for mischief because Pakistan needed missiles and was willing to trade nuclear weapons technology and equipment.

On October 9, 2006, North Korea announced that it had exploded a nuclear weapon underground. American, Chinese, and Russian scientists detected an explosion equivalent to five hundred tons of TNT, and a few days later, a surveillance airplane detected radioactivity in the atmosphere. The magnitude was very small for a nuclear weapon, which would be 10 to 20 times larger, and the radioactivity was slight. Apparently it was a dud. It takes excellent engineering, not just a lump of plutonium, to achieve a yield above a couple of kilotons. It also takes high-quality components like the detonator, the explosives to compress the sphere, and so forth. Even though the test was a dud, however, it seemed likely that North Korea could develop a successful atom bomb within a few years. Furthermore its missiles will improve. A few bombs could destroy South Korea or cripple Japan. These two countries lack nuclear capability and depend on the United States for deterrence. U.S. nuclear weapons could destroy North Korea in a few minutes. China has its own nuclear forces and seems inclined to retaliate if attacked. An attack on the north might be limited or might ignite a bigger war.

Faced with this extreme danger, the United States, along with Japan, South Korea, China and Russia, decided to go back to the bargaining table. The six parties revived the previous agreement of supplying 500,000 tons of free oil annually in exchange for North Korea ending its nuclear weapons program. In essence, it was the agreed framework of 1994, only now North Korea had several bombs in its arsenal. It seemed to be the best deal the United States and the others could get.

ANTIMISSILE DEFENSE: STAR WARS REDUX

Reagan's SDI proposal did not disappear, but slumbered quietly during the Bush Senior administration. Clinton tried to avoid the issue. When during the 1996 election, the Republican nominee, Bob Dole, tried to make missile defense an issue, Clinton neutralized him by advocating his own version, by now known as national missile defense. To defuse the issue, Clinton appointed a commission chaired by the former (and future) Republican secretary of defense, Donald Rumsfeld. The commission recommended proceeding with the missile defense scheme. For one thing,

Clinton was afraid that otherwise the Republicans would use it in the 2000 campaign as an issue against his vice president and designated successor, Al Gore. Technologically, the system faced problems. Of the three tests performed, the defending missile only succeeded in shooting down the incoming missile once. Critics complained that the tests were rigged in favor of the system by using old, slow missiles as targets. In the one success, the defending missile actually hit the target because it zeroed in on a dummy balloon nearby. When a test in July 2000 was a failure, Clinton indefinitely postponed moving forward, leaving Gore off the hook for the rest of the campaign.

The exact nature of antimissile defense remains fluid, which suits its advocates fine. Those who want a full-scale, global scheme defending against an all-out Russian attack recognize that advocating the full scope would frighten the Russians and the Chinese, and European and Pacific allies, as well as the American taxpayers, hence they do not articulate the big version. The lowest level is designed to protect against small rogue states like Iran and North Korea. In fact this is not a "national" defense at all, but a local one. The American defensive missiles would be based in South Korea, Persian Gulf countries, and on naval ships. This presumes that a small system countering the North Koreans would not agitate the Chinese. The next level would be aimed against the Chinese, who actually are known to possess 20 ICBMs. Presumably this midsized system would not agitate the Russians. The biggest system, often called "multilayered," would be global to protect the United States against Russian attack over the North Pole. If the Russians ever launched an all-out attack and the United States successfully defended (that is, shot down 100 percent), the result would be the explosion of 5,000 nuclear bombs generating radioactivity that would kill nearly the entire population of the world and lead to nuclear winter.

For the United States to move forward with the national missile defense would lead to several problems. First, it would appear to the rest of the world as aggressive bullying. Second, militarizing space could hurt Americans more than Russians because the Americans depend more on satellites for communications (civilian and military) and for intelligence gathering though photography. European allies worry that a national missile defense might cause the United States to go it alone, without regard for them. In the 2000 election Governor Bush promised to build the missile defense, which brought him enthusiastic support from conservative Republicans.

After the September 11 attacks on the World Trade Center and the Pentagon, the debate on missile defense became conflated with the issue of homeland security from terrorists. Advocates argued that it would protect against rogue states like North Korea that did not participate in civilized interaction with the rest of the world and were unpredictable. In his State of the Union address to Congress, President Bush warned against an Axis of Evil (North Korea, Iran, and Iraq) that threatened aggression with weapons of mass destruction. Bush defined these as nuclear bombs, poison gas, and disease pathogens. Of the three, only North Korea was then close to building a nuclear weapon. Although Iraq had tried to build an atom bomb at the Osirak reactor near Baghdad, the Israeli Air Force attacked and destroyed the facility in 1981. Searches after the war showed that the Iraqis were not able to recover from this, and they abandoned their nuclear project. Many wondered why the president defined those three countries as the Axis of Evil since they had not supported the Al Qaeda organization that had attacked on September 11. Indeed, 15 of the 19 airplane hijackers came from Saudi Arabia. The Taliban regime in Afghanistan that had supported Al Qaeda was quickly defeated and replaced. Sudan, which had supported Al Qaeda earlier, was ignored. Focusing on the Axis of Evil ignored India, Pakistan, and China, all of which had nuclear missiles.

Bush's other response was to announce that the anti-ballistic system was ready for deployment to protect against North Korea. This was a greatly scaled-down version with six interceptor missiles stationed in Alaska and four in California. It was renamed Ground-Based Missile Defense, the other (future) components being air and naval. President Bush deployed the system even though the tests were mostly failures. Two were successful however. Critics pointed out that even these tests were biased. The cost was estimated at $50 billion over five years.

CONCLUSION

Not only is nuclear war the most apocalyptic danger facing the world, it is also the one that draws the most literary references to the Bible. The scientists at the Alamogordo test said they had witnessed the potential destruction of the world. The *Bulletin of the Atomic Scientists* features the Doomsday Clock, recalling medieval religious terminology of the Apocalypse. A small war could kill 20 million, and a big war setting off nuclear winter would kill billions.

The danger of world destruction did not come into popular discussion until nearly a decade after Hiroshima for two reasons. The first was secrecy. Although Americans watched news films and read stories of bomb tests, they did not have the total picture of the danger. Then in 1954 Eisenhower and Dulles addressed the issue publicly, frightening many citizens with the plan of massive retaliation. A second reason was that the government itself, even at the highest levels, had not recognized the enormity. It had just been building A-bombs, one after another, until suddenly it had 2,000 of them. Like the top civilian leadership, the military had not done sophisticated planning. It simply intended to drop as many as possible on the Soviet Union in the Sunday Punch. To be fair to Eisenhower, in order to put safeguards on the weapons, he needed the cooperation of the Soviets, and this could not come until after the death of Joseph Stalin and the transition to stable leadership under Khrushchev. Stability in the Kremlin allowed negotiation of the Test Ban Treaty, the Non-Proliferation Treaty, the ABM Treaty, and SALT I. In many ways complex treaties like these are like plans.

Dramatizing the danger of nuclear winter did not spark a response from the Reagan administration. It was merely ignored. The same can be said for citizen demonstrations for a nuclear freeze at the same time. Indeed the nuclear policy of Ronald Reagan was the Strategic Defense Initiative: Star Wars. Had this ever actually been planned for it would have been extremely complex. Since the furor over SDI died down, little comprehensive planning has occurred. Bill Clinton let a shadow of the scheme continue because it seemed an easy way to avoid Republican criticism, and George W. Bush came to believe in the program and hence actually deployed a miniature version.

NOTES

1. Henry A. Kissinger, *Nuclear Weapons and Foreign Policy* (New York: Council on Foreign Relations, Harper Brothers, 1957), p. 70.

2. Peter Goodwin, *Nuclear War: the Facts on Our Survival* (New York: Rutledge, 1981), p. 8.

3. Jeannie Peterson, ed., *Nuclear War: the Aftermath* (Oxford, Pergamon Press, 1983), based on a special issue of *Ambio* XI (1982), p. v.

4. Carl Sagan and Richard Turco, *A Path Where No Man Thought: Nuclear Winter and the End of the Arms Race* (New York: Random House, 1990), p. 7.

5. Richard G. Hewlett and Jack M. Holl, *Atoms for Peace and War* (Berkeley: University of California Press, 1989), pp. 154–155.

6. Office of the Federal Register, National Archives and Records Service, *Public Papers of the Presidents, Dwight D. Eisenhower, 1960,* pp. 1035–1040.

7. Herman Kahn, *On Thermonuclear War* (Princeton: Princeton University Press, 1960), pp. 357–371.

8. Bernard J. Firestone, *The Quest for Nuclear Stability* (Westport, CT: Greenwood, 1982), p. 49.

9. Roger Hillsman, *The Cuban Missile Crisis* (Westport, CT: Praeger, 1996), p. 15.

10. Ibid., p. 97.

11. Carl Kaysen, "The Limited Test Ban Treaty of 1963," in *John F. Kennedy and Europe* (Baton Rouge: Louisiana State University Press, 1999), p.112.

12. John Newhouse, *War and Peace in the Nuclear Age* (New York: Knopf, 1989), p. 137.

13. Ibid., p. 271.

14. Joseph S. Nye, "The Superpowers and the Non-proliferation Treaty," in *Superpower Arms Control,* ed. Albert Carnesale and Richard N. Haass (Cambridge, MA: Ballinger, 1987), p. 175.

15. Fen Osler Hampson, "SALT I: Interim Agreement and ABM Treaty," in *Superpower Arms Control,* ed. Carnesale and Haass, p. 71.

16. Newhouse, *War and Peace in the Nuclear Age,* p. 337.

17. Thomas R. Rochon and David S. Meyer, "The Nuclear Freeze in Theory and Action," in *Coalitions and Political Movements,* ed. Rochon and Meyer (Boulder, CO: Lynne Rienner, 1997), p. 6.

18. David S. Meyer, *A Winter of Discontent: The Nuclear Freeze and American Politics* (New York: Praeger, 1990), pp. 123–125.

19. Rochon and Meyer, *Coalitions and Political Movements,* p. 8.

20. Richard P. Turco, Owen B. Toon, Thomas P. Ackerman, James Pollack, and Carl Sagan, "Nuclear Winter: Global Consequences of Multiple Nuclear Explosions," *Science* 222 (1983): 1283–1297.

21. Sagan and Turco, *A Path Where No Man Thought.*

22. Ronald Reagan, "Peace and National Security," *Weekly Compilation of Presidential Documents* 19 (March 23, 1983): 443.

23. Ibid.

24. Donald R. Baucom, *The Origins of SDI 1944–1983* (Lawrence: University of Kansas Press, 1992), p. 146.

25. Newhouse, *War and Peace,* pp. 360, 363.

26. Oberdorfer, *From the Cold War to a New Era* (Baltimore: Johns Hopkins University Press, 1998), pp. 196–198; Frances Fitzgerald, *Way Out There in the Blue; Reagan, Star Wars and the End of the Cold War* (New York: Simon and Schuster, 2000), p. 363.

27. Oberdorfer, *From the Cold War to a New Era,* pp. 259–260.

28. Fitzgerald, *Way Out There in the Blue,* p. 427.

29. Oberdorfer, *From the Cold War to a New Era,* p. 381.

30. Bruce G. Blair, Harold A. Feiveson, and Frank N. von Hippel, "Taking Nuclear Weapons Off Hair-Trigger Alert," *Scientific American* (November 1997), cited as von Hippel.

31. James Goodby, "Transparency and Irreversibility in Nuclear Warhead Dismantlement," in *The Nuclear Turning Point,* ed. Harold Feiveson (Washington, DC: Brookings, 1999), pp. 181, 184, 177.

32. PBS, "Frontline: Loose Nukes," *www.pbs.org/wgbh/pages/frontline/shows/nukes/.*

33. Von Hippel, "Taking Nuclear Weapons."

34. Ibid.; Bruce G. Blair, "De-alerting Strategic Nuclear Forces," in *The Nuclear Turning Point,* ed. Feiveson, pp. 104–105.

35. Robert S. Norris and Hans M. Kristensen, "Nuclear Notebook: Global Nuclear Stockpiles, 1945–2006," *Bulletin of the Atomic Scientists* 62 (2006): 64–66.

36. Akbar Hashemi-Rafsanjani, "Qods Day Speech (Jerusalem Day)," Voice of the Islamic Republic of Iran, in Persian. Translated by BBC Worldwide Monitoring December 14, 2001.

— 6 —

Global Warming:
International Scientific Planning

The first alarm about global warming came in 1896 when the Swedish chemist Svante Arrhenius warned that "we are evaporating our coal mines into the air."[1] In 1972 the Club of Rome cautioned that the carbon dioxide in the atmosphere was increasing exponentially.[2] Testifying to Congress in the midst of a record-breaking 1988 heat wave, NASA scientist James Hansen explained that global warming due to carbon dioxide caused heat waves and droughts. As temperatures warmed in Antarctica, icebergs as big as Rhode Island broke off the Larsen Ice Shelf. Melting of glaciers on the mainland and of the Greenland icecap could cause the sea level to rise, threatening coastal cities around the world. Also in Antarctica, scientists discovered that the ozone layer of the atmosphere was thinning and for a few months each year creating a hole.

Like the main plot and a secondary plot in a drama, the saga of global warming twists around the saga of the ozone hole. The revelation of the ozone hole problem and its solution in the form of the Montreal Protocol became a model. The dangers of warming caused by greenhouse gases are far greater than the danger of the ozone layer thinning, but the scientific evidence of the hole came suddenly and unambiguously, and national governments acted promptly to correct the problem. The plot for warming has been slower, more complex, and more ambiguous.

THE GREENHOUSE EFFECT

The 1896 prophecy of Arrhenius, that the earth's climate was warming due to carbon dioxide, must rate at the top of apocalyptic forecasting. In his article that year in the *Philosophical Magazine,* which he published in English for the "benefit of the Anglo-Saxons," Arrhenius explained that increased carbon dioxide in the atmosphere caused heating and that burning coal was the chief contributor, and he went on to calculate that doubling the amount in the atmosphere would cause a temperature rise of 11 degrees F, not much higher than the present calculation of 5–9 degrees. His stimulus came from the Swedish Physics Society debate about the causes of the Ice Age, which still lingered in northern Scandinavia. Modern instruments made it possible to measure the amount of atmospheric gases, and Arctic exploration had expanded their scope. Arrhenius himself had served as hydrographer on an expedition to Spitsbergen in the Arctic Ocean. Arrhenius speculated on the future effects, noted the effect of water vapor, pinpointed coal as the culprit, and painstakingly computed the results for five scenarios at 14 latitudes from 70 degrees north to 60 degrees south during each season.[3] He compared the phenomenon to a hothouse. In contrast to the present view, Arrhenius believed that global warming would be good because it would expand agriculture into northern lands like Sweden.

Almost immediately, climatologists took up Arrhenius's theory, admiring his scope and detailed calculations. Thomas Chamberlin of the University of Chicago amplified it with alternative explanations for changes in carbon dioxide due to changes in the relative amount of land and sea areas, which resulted in limestone being exposed or being created. Thus, by the turn of the century, the basic concept and measurements of the theory were in place.[4] During the 1930s, while Europe was experiencing a series of very hot summers, the British coal engineer George Callendar compiled temperature data from around the world, including readings of the ocean taken by sea captains. The data showed a smooth upward trend, which he hypothesized was due to carbon dioxide. Ironically, after he published his results in 1938, the climate of the northern hemisphere began to cool. However, most meteorologists who read his theory believed that carbon dioxide would not be a problem because the ocean would absorb any excess amounts, locking it away in sediments at the bottom.[5]

In 1957 Roger Revelle and Hans Suess came to the contrary conclusion, after they determined that sea water could absorb only a limited amount of carbon dioxide. Moreover, widespread burning of fossil fuels was adding excessive amounts. They wrote that "human beings are now carrying out a large-scale geophysical experiment of a kind that could not have happened in the past nor be reproduced in the future."[6]

Charles Keeling measured trace gases on the 11,000-foot summit of Mauna Loa in Hawaii, where the remote location and extreme altitude removed local effects. His manometer revealed seasonal fluctuations, low in the spring and summer as trees and plants in the northern hemisphere leafed out and grew trunks, stems, and roots. In the fall and winter, it became high as the plants released their carbon dioxide when their leaves fell and decayed. Each year the same cycle occurred with the seasons. But when Keeling examined the longer term of several years, he discovered a gradual increase. In 1958 the level was 315 parts per million, and on average, it increased more than 1 ppm annually. Keeling compared this to some antique air samples from nearly a century before when industrialization was much less, which were only 280 ppm. This meant that in the 70 years since then factories and automobiles had increased the amount of carbon dioxide in the atmosphere by 12 percent. Although the Keeling Curve, as it became known, pointed toward a major threat, the earth was again in a period of cooling, which lasted 35 years. Thus the facts contradicted the theory.[7]

Carbon dioxide is not the only greenhouse gas; others are methane, nitrous oxide, and CFCs, whose combined impact is roughly equal to carbon dioxide. Methane is the natural product of bacterial decay in the absence of oxygen. The 1.3 billion cows in the world are the biggest single source. Methane is a component of natural gas burned in houses and factories and is present in many coal deposits. Large amounts are locked up in sediments on the bottom of the ocean. The amount doubled in the twentieth century. Nitrous oxide, another greenhouse gas, comes from bacteria in the soil. Farmers use nitrogen compounds as fertilizer, whence it passes into the air. Its levels have increased 8 percent in the past century. CFCs—chlorofluorocarbons—are not natural, but were invented in the 1930s by General Motors for automobile cooling systems and were soon used for refrigerators. The Du Pont company sold them under the name Freon. In addition to being a greenhouse gas, CFCs drift up to the stratosphere where they destroy the ozone layer.

Timeline for Global Warming

1896   Arrhenius discovers global warming caused by CO_2 and calculates its effect on glaciers.

1931   CFCs invented.

1938   George Callendar publishes article showing that the temperature of seawater is rising.

1957   Roger Revelle and Hans Suess publish article on global warming.

1958   Keeling Curve shows rising temperatures.

1965   President's Science Advisory Committee reports problem of global warming.

1969   Congress holds hearings about global cooling. More hearings held in 1972 and 1974.

1970   Beginning of debate over supersonic transport damage to ozone.

1974   Sherwood Roland and Mario Molina propose theory of CFC damage the ozone layer.

1977   EPA and FDA ban most aerosol spray cans.

1985   Ozone hole discovered above Antarctica. Workshop in Villach, Austria, concludes warming is "highly probable."

1987   CFC limits negotiated to protect the ozone layer in the Montreal Protocol. Hottest year on record.

1988   Toronto Conference proposes a 20 percent reduction in CO_2 by the first world.
Record-breaking summer heat in North America; second hottest for entire world.

1990   Second World Climate Conference in Geneva; hottest year on record.

1992   Framework Convention on Climate Change signed at Earth Summit in Rio de Janeiro.

1995   First Conference of the Parties issues the Berlin Mandate.

1997   Kyoto Protocol signed; hottest year on record.

1998   Hottest year on record.

2001   Pres. George W. Bush condemns the Kyoto Protocol.

2005   Kyoto Protocol enters into force after Russia ratifies it; hottest year on record.

THE SUPERSONIC TRANSPORT

Ozone was the first stratospheric issue to reach the American political agenda. In 1962 the French and British governments signed an agreement to develop a supersonic transport airplane, named the Concorde. These aircraft were designed to fly at an altitude as high as 70,000 feet in the stratosphere, traveling from London and Paris to New York or Washington in three hours. The Soviet Union planned its own SST, the TU 144. These countries sought both the prestige and sales to foreign markets. At the time, 85 percent of the airplanes sold worldwide were American.[8]

Environmentalists soon opposed the project, but not because of the ozone danger. They objected to the SST's sonic boom, its noise during takeoff, and its ground-level pollution. People under the path of the SST heard a loud boom as the plane flew overhead caused by the shock wave when its speed was greater than sound. The boom afflicted the entire flight, not just its takeoff and landing, which were also noisy. The noise of an SST taking off was four times as loud as a conventional jet. The exhaust spewed out nitrous oxide. For three years the Federal Aviation Administration had conducted a rather cynical experiment by bombarding residents of Oklahoma City with booms from military jets to see if citizens would grow accustomed to the noise. In 1967 William A. Shurcliff, a physicist at Harvard University, organized the Citizens League Against the Sonic Boom. At first the league consisted only of Shurcliff, his sister, his son, and one friend, but by 1970 it had 5,000 members. It put out a stream of press releases, paid advertisements, and letters to members of Congress. The FAA responded with its own press releases, arguing that the United States needed to build its SST to counter the Concorde and the Soviet TU 144.[9]

Although Congress had approved funding with little opposition in 1969, the debate the next year was heated. Environmentalists mobilized under the inspiration of David Brower, for 16 years the executive director of the Sierra Club and then founder of Friends of the Earth. The coalition against the SST ranged from the Friends of the Earth to the National Taxpayers Union to the Oil Workers Union. Shurcliff's Citizens League against the Sonic Boom affiliated with it. The coalition's two main objections to the SST were the boom and the expense. Ozone was not considered important. In the Senate, William Proxmire of Wisconsin led the opposition. His first problem was that the appropriation went before the Subcommittee on Transportation of the Appropriations Committee, chaired by John Stennis, who also chaired the full Committee on Armed Services. Senator Stennis, one of the old great powers in Congress, treated the SST more like a military project than a civilian one, keeping information secret and hinting that the air force would benefit.

To gain a forum, Proxmire decided to also hold hearings before the Joint Subcommittee on Economy in Government, which he chaired. This minor subcommittee would provide a platform for critics, and Proxmire was a master at getting publicity. Most witnesses concentrated on the waste of money, but a few spoke of environmental risk. On the third day, Russell Train testified. Congress had passed the National Environmental Policy Act only five months earlier, which required environmental impact statements and established the Council on Environmental Quality. Train was its first chairman, and in that period before the establishment of the Environmental Protection Agency, he was the Nixon administration's highest ranking environmental official.

Train testified that a fleet of five hundred SSTs might disrupt weather patterns if they flew at an altitude of 60,000–70,000 feet by emitting water, carbon dioxide, and nitrogen oxides into the stratosphere. He briefly mentioned the risk of destroying ozone. Train's warnings were similar to those found in an MIT study published contemporaneously, titled "The Study of Critical Environmental Problems." Having Russell Train give the same information validated it.[10] Shortly thereafter on the other side of the Capitol, the House of Representatives voted 107–86 to appropriate $290 million for the SST. Although the opponents lost, they were elated that they had won more votes than in previous years. Over the summer a number of senators announced that they had changed their minds and no longer supported the SST. The Senate Subcommittee on Transportation of the Appropriations Committee began its own hearings in late August. Department of Transportation officials testified in favor, and the CEQ officials testified against. As in the House hearing, ozone was mentioned only as a secondary risk. The Coalition Against the SST solicited letters from prominent economists, who overwhelmingly considered it a giant waste of the taxpayers' money. To ensure maximum publicity, Senator Proxmire read them into the *Congressional Record*.

The general election coming up in November caused a number of senators to want to postpone the appropriations bill. Proxmire skillfully fought with his few allies and limited resources. He was a master at publicity and parliamentary procedure. His strongest opponent was the senator from Washington state—where Boeing was scheduled to manufacture the airplane—Henry Jackson. During his 42-year career in Congress, Jackson was one of the best friends the environment ever had. He sponsored NEPA, the Wilderness Act, the Wild and Scenic Rivers Act, and the Alaskan Lands Act. His love of nature was only terrestrial, however, for he favored building dams on the Columbia River and manufacturing aircraft at the Boeing factories in Seattle. During the following months, Proxmire was able to filibuster, so the Senate postponed the vote and eventually abandoned the bill.

The congressional saga had its high drama and moments of humor. Proxmire called a new witness to testify: Professor James McDonald of the University of Arizona, who had studied ozone depletion by SSTs. McDonald concluded that a fleet of five hundred SSTs would reduce the ozone layer enough to cause 10,000 additional cases of skin cancer per year. This was the most explicit health danger predicted so far, and the first time ozone emerged from being a secondary issue. McDonald appeared before the House subcommittee, making a clear and persuasive case. Then to the surprise of Proxmire and the coalition, a Republican representative asked a question that nearly destroyed his scientific credibility: Aren't you the scientist who believes in flying saucers?[11] Indeed, four years previously, McDonald had testified before Congress that UFOs had caused a massive electrical blackout on the East Coast. In a desperate attempt to salvage the testimony, Proxmire asked other scientists to comment on McDonald's claim. Fortunately two prominent scientists announced that they believed McDonald's study was good science, and one said that in fact, the fleet would cause 23,000 to 103,000 cases of cancer.[12]

Completion of the first Concordes reactivated the SST issue in 1975, when Air France and British Airways applied for permission to fly to New York and Washington. Because it would be a "significant Federal action," NEPA required an environmental impact statement. Two items made the Concorde more acceptable than five years earlier. The airlines promised not to fly at supersonic speed over land, eliminating the sonic boom. The number of planes would be few; not the fleet of five hundred predicted earlier, and the Concorde would fly at 55,000 feet, lower than the 70,000 feet predicted earlier. The FAA concluded the Concorde would cause only two hundred additional cases of skin cancer in more than 30 years. Its environmental impact statement did find it would cause problems of excessive noise and emissions on takeoff.

The biggest problem for the Concorde, however, was economic, just as predicted by Proxmire, the coalition, and two dozen leading economists. The price of a ticket was at least twice as expensive as a regular jet, and few passengers considered saving two hours worth the expense. Air France and British Airways were unable to find any other profitable routes for the airplane, terminating service to Singapore and Buenos Aires after a few years. No other airlines purchased Concordes. Once on scheduled service, they proved to be unreliable and expensive. In the year 2000, a crash on takeoff from Paris killed 113 and kept the aircraft out of service for a year. When flights resumed, passenger demand was cut in half, and in 2003 Air France and British Airways ended service. The Soviet TU 144 did even worse. A prototype crashed at the Paris Air Show in 1973; commercial service was not successful, and today the remaining aircraft are in a museum.

CFCS AND THE OZONE LAYER

At the same time as concern about protecting the ozone layer from the SST was dying down because of the economic failure of the aircraft, two chemists recognized danger from a different direction. Sherwood Rowland and Mario Molina discovered that CFCs from refrigerators, air conditioners, and aerosol spray cans drift up to the stratosphere where they act as catalysts to destroy the ozone. Ultraviolet radiation from the sun decomposes the CFC molecule, releasing a chlorine atom, which reacts as a catalyst to split apart thousands of ozone molecules. The potential damage was great, but would take many decades to develop. It could cause 150 million new cases of skin cancer resulting in 3 million deaths in the United States. The additional ultraviolet light could cause 18 million additional eye cataract cases. Two hundred plant species, including peas, cabbage, melons, and cotton, are sensitive to ultraviolet harm. Evidence suggested that it could disrupt the aquatic food chain based on plankton floating near the ocean surface.[13]

The Rowland and Molina article, published in 1974, stimulated debate. Industry manufactured 500 million spray cans annually, and the average family had 40 or 50 in its house. About half of the cans used CFCs, particularly hair sprays and deodorants. They had the advantages of being odorless, nontoxic, nonflammable, and chemically inert. The second most common use was air conditioners and refrigerators, and the third was blowing foam for Styrofoam cups and home insulation. Worldwide production of CFCs totaled 800,000 tons annually, half in the United States.

The American public quickly grew alarmed. *Harpers* magazine warned readers that they could push their own button of destruction and warned them to "listen for that little whisper of doom." The *Philadelphia Inquirer* proclaimed that "the earth will end—not with a whimper, but with a quiet pfft."[14] Industry mobilized a sophisticated public relations defense coordinated by the Aerosol Education Bureau. An allied Council on Atmospheric Science organized industry scientists and spokesmen to testify at congressional and state hearings and give public speeches. In fact, no members of the council itself were actually scientists. Du Pont, the world's largest manufacturer of CFCs, skillfully used advertisements, briefings, and press releases. Industry broke ranks, however, when companies that manufactured aerosol cans with no CFCs began to advertise the fact. The Du Pont company began research to find substitutes, but once Ronald Reagan won the White House, with his promise to ease regulations, it abandoned the effort for six years until the initial pro-business policies of the new administration has been moderated and later fresh evidence from Antarctica reemphasized the problem.[15]

Congress moved promptly to investigate. The National Academy of Sciences began a study. So did the Federal Interagency Task Force on the Inadvertent Modification of the Stratosphere, under the leadership of the NASA. The Natural Resources Defense Council, an environmental group, petitioned the Consumer Product Safety Commission to ban CFCs in household cleaning supplies and refrigerators. People called on the Food and Drug Administration to ban them in hair sprays and deodorants. Within months, the Task Force reported that CFCs were a serious cause for concern and suggested that the chemicals be regulated. Industry defended its position by latching onto any scientific item that countered the CFC theory. Corporate public relations departments denied that CFCs reached the upper atmosphere or that they damaged the ozone layer. They argued that no action needed to be taken at once because the delay of a few years would be insignificant.[16]

In spite of industry opposition, the regulatory agencies were persuaded that CFC aerosols posed a danger to the ozone layer. The Food and Drug Administration determined that cosmetics using them should be labeled. The Consumer Product Safety Commission acceded to the Natural Resources Defense Council petition to ban CFCs in spray cans, and EPA announced its intention to ban the aerosols. The three agencies jointly announced a ban on the aerosols fully effective in 1979. Congress amended the Clean Air Act to give EPA explicit authority. Once the ban was announced, industry was able to find substitutes comparatively easily. CFCs could still be manufactured for nonaerosol use like refrigeration and air conditioning. Regulations provided for recovery of the chemicals when the equipment was junked.

Meanwhile, defense experts were investigating whether the explosion of nuclear weapons would damage the ozone layer. Evidence suggested that a major war would destroy 50 to 75 percent of the ozone layer. A person would get a blistering sunburn in 10 minutes, and the harm to health would be greater than that from radioactive fallout.[17]

The American ban covered half of the world production. The European Community response was weaker. Although West Germany banned the aerosols, France, Italy, and Britain refused, due to pressure from their industries. The Japanese also resisted regulating CFCs. As each year passed, however, the dangers became more apparent. The UN Environmental Program convened representatives of 24 countries in Stockholm to discuss the technical and legal issues in a future treaty. In 1985 43 countries gathered in Vienna under the auspices of the UN Environmental Program to draft a treaty, known as the Vienna Convention. Its provisions were weak and vague, so the Europeans were willing to sign. The parties agreed to conduct research, observe the atmosphere, conduct workshops, and "take appropriate measures" (which were not defined). The Vienna Convention provided for a Conference of the Parties to meet on a regular basis and established a secretariat.[18]

The next step was to draft a protocol, that is, a treaty within the framework of the Vienna Convention, to put teeth into implementation. In September 1987 the parties convened in Montreal, where they ratified agreements worked out in the prior year and a half and negotiated a few remaining issues. Montreal was a logical site since the Canadians had been leaders in the movement to protect the atmosphere. Moreover, as a middle-sized power in international relations, Canada had found a niche in environmental diplomacy. It could provide leadership and focus attention here when the great powers were preoccupied with economic strategy and potential wars.

One issue was the base year. Most countries wanted 1986, in which they were bargaining, but the European Community insisted on 1990, supposedly because it would be hard to gather statistics for a while, but actually because its companies intended to increase production temporarily for the next few years. A second issue was the implementation schedule, on which the parties compromised with a freeze, then reductions of 20 and 30 percent in the following years. A third issue was special privileges for underdeveloped countries, who argued that they deserved this because they were poor. The compromise was to permit a 10-year grace period and to give the third-world countries higher quotas.[19] The three provisions became a model for the later negotiations on global warming.

The European Community also sought special privileges. It argued that because the new Single Europe Act would go into effect in 1992, the 12 member-countries should be treated as a unity, much like the 50 United States. This would allow the community to report emissions collectively, which would mean the large reductions already made by Germany could offset high levels by other countries. At the same time as it argued that Europe was a unity, however, it insisted that all 12 members should be entitled to a vote.[20] This also became a model for the global warming negotiations.

Within the U.S. government, a bargaining position needed to be developed to accommodate diverse interests. The State Department led a committee of staff representatives from EPA, the Commerce and Energy departments, NASA, OMB, and the president's Special Trade Representative. Early on it decided to abandon its 1985 goal of a worldwide ban on aerosols because of intense European opposition. Instead it would propose limits. This was still not fully acceptable to more conservative, pro-business advisors close to President Reagan. At the time of the Vienna conference, the assistant secretary of state for economic affairs, Allen Wallis, had tried, at the last minute, to block the U.S. signing because he feared it would lead to international regulation of business.[21] Within the State Department, the leadership came from a different division, the office of the assistant secretary for environmental affairs, but all department factions were required to be reconciled. The effective coordination at the staff level avoided bringing the controversies to the attention of the high levels at the White House. EPA and the scientific agencies dominated the decision making. The technical people avoided alarming the political people.

Contemporaneously with the ozone negotiations came provocative news from Antarctica. Two years earlier British scientists had first published their discovery of the ozone hole there in 1985.[22] Now as the diplomats were finalizing the protocol, an expedition of 60 scientists from the United States, Britain, and other countries were in Antarctica trying to get definitive information. Their preliminary data released a few months later gave more evidence pointing to CFCs as the cause. Further research discovered that the Antarctic was not the only vulnerable location. The ozone layer was thinning over the northern hemisphere, and a hole was forming over the Arctic.

At a Conference of the Parties in 1990 meeting in London, the industrial countries agreed to give money to the third-world countries to finance compliance.[23] Initially the Multilateral Fund was $240 million, which increased to $1 billion. This was the first time the industrial and developing countries had set up a fund as an incentive to environmental cooperation, and with only a few rough spots, it functioned moderately well. After the initial enthusiasm, the European countries realized that the fund would be expensive and tried to reduce their commitments, but eventually they came to accept its necessity. The World Bank guided the handouts.

At the same time, unscientific attacks challenged the phaseout. Congressman Robert Dornan inserted an article into the *Congressional Record* characterizing it as "the most scientifically baseless, politically oppressive, morally bankrupt, economically destructive environmental farce."[24] Congressman Tom Delay introduced a bill to repeal American implementation of the treaty.[25] Both representatives were conservative Republicans. The radio host Rush Limbaugh described the phaseout as "balderdash. Poppycock."[26] S. Fred Singer, an atmospheric physicist at the University of Virginia known for his iconoclasm, asserted in congressional testimony that "I want to state clearly that there is no scientific consensus on ozone depletion or its consequences....The 1987 Montreal Protocol was negotiated without adequate concern for scientific evidence."[27] He went on to conclude that

currently available scientific evidence does not support a ban on the production of chlorofluorocarbons (CFCs or freons), halons, and especially methyl bromide. There certainly is no justification for the accelerated phase out of CFCs, which was instituted in 1992 on nothing more than a highly questionable and widely criticized NASA press conference. Yet because of the absence of full scientific debate of the evidence, relying instead on unproven theories, we now have an international treaty that will conservatively cost the U.S. economy some $100 billion.[28]

The history of the CFC-ozone depletion issue is rife with examples of the breakdown of scientific integrity: selective use of data, faulty application of statistics, disregard of contrary evidence, and other scientific distortions. The policy before and since the Montreal Protocol has been driven by wild and irresponsible scare stories: EPA's estimate of millions of additional skin cancer deaths, damage to immune systems, blind sheep in Chile, the world-wide disappearance of frogs, plankton death, and the collapse of agriculture and ecosystems.[29]

In spite of the problems of implementation and attacks from the right wing, scientific and governmental authority continued to support the phase-out. The Montreal Protocol stands as one of the most successful international treaties. Agreement came fairly quickly with minimal conflict. It has accomplished most of its goals of ending the danger CFCs pose to the ozone layer.

CONSENSUS ABOUT THE GREENHOUSE EFFECT

Unlike the smooth resolution of the ozone issue, the warming of the earth's atmosphere due to burning fossil fuels was more difficult. Until 1985, most scientists did not accept the greenhouse theory. For the most part, they reserved judgment, and some believed the earth was cooling. Since 1940 the trend had been cooling, and especially from 1958 to 1977.[30] In 1969, 1972, and 1974 Congress held hearings on the threat of cooling. The immediate stimulus came from Department of Defense proposals to use weather modification as a weapon. At the time, B-52 bombers over Indochina were dropping Agent Orange herbicide to defoliate the rain forest so the Viet Cong and North Vietnamese troops could not hide. Airborne cloud seeding could stimulate rain in one region and take it away from another region. Other testimony told how the northern hemisphere had cooled one degree F in the preceding three decades and predicted more cooling. Most ominously, it forecast that slightly more cooling would tip the balance and plunge the planet into a new Ice Age. The National Academy of Sciences issued a report in 1974 suggesting that a new Ice Age could begin within a hundred years, and the Central Intelligence Agency predicted that the climate would cool. After reviewing the scientific literature, the CIA concluded that most climatologists believed that the weather in the early twentieth century had been abnormally warm and would return to a lower, normal temperature. Evidence was clear: "Arctic ice is increasing; the English growing season is now a week shorter than it was in 1940; and other evidence from Iceland indicate that the last four decades have been the most abnormally warm period in the last 100 years."[31]

The cooling forecast had strong evidence. The previous century had been one of the warmest eras in geologic history. The earth warmed up from a quarter million years of glaciation and cold a mere 10,000 years ago, and soon after this, homo sapiens began farming and living in villages. It was the beginning of civilization. The warmest era was about 1100 to 1250, known as the medieval climate optimum or the medieval warm period. More recently, from 1430 to 1820, the northern hemisphere endured a little ice age. Crops in Europe failed, the Vikings lost contact with their colonies in Greenland, and the Atlantic ice pack extended to Scotland.[32] The very coldest years were the turn of the nineteenth century. In America, the Boston and New York harbors froze regularly, and George Washington and the Continental Army suffered in Valley Forge. New Englanders called 1816 the "year with no summer." Snow fell in July and crops perished from frost. Then the climate reversed, and the period from 1880 to 1940 was warmer. Arrhenius and Callendar conducted their research during this time. After this the temperatures fell again.[33] The cooling trend ended by 1977, and the temperatures warmed. Although climatologists claim that they look only at the long term, the short-term reversal made it easier for them to make their case. Improved technology contributed too. Bigger computers supported a three-dimensional general climate model, and weather satellites provided volumes of data. Meteorological models are the most complex of any computer programs.

Within two days of each other in 1983, the federal government issued two contradictory reports on global warming. The National Academy of Sciences recommended no action, and EPA recommended immediate action. In fact, both reports agreed that carbon dioxide in the atmosphere had increased, but the academy concluded, "we find the CO_2 issue reason for concern, but not panic," and that "increased CO_2 may also bring benefits."[34] Both Revelle and Keeling served on the panel. EPA, in contrast, concluded that carbon dioxide, combined with other greenhouse gases, would bring a crisis by the year 2030. "As a result agricultural conditions will be significantly altered, environmental and economic systems potentially disrupted, and political institutions stressed."[35] The immediate impact of the two reports was nil. The Reagan administration was only a few months past the political disaster at EPA when its director, Anne Gorsuch Burford, had resigned in disgrace. Top political appointees were found to be manipulating information, disobeying the law, and flouting the requests of key congressmen. One went to jail for corruption.

The evidence on global warming accumulated during the next few years. In 1985 the United Nations sponsored a technical workshop in Villach,

Austria, which concluded that the accumulation of greenhouses gases made global warming "highly probable."[36] In addition to the problem of carbon dioxide, research published that year showed that three other greenhouse gases (methane, nitrous oxide, and CFCs) contributed an equal danger. In other words, the problem was twice as great as with carbon dioxide alone. The participants stepped beyond pure science to recommend that it was time to address this in terms of policy.[37] Over the next three years the potential danger of global warming gained attention. Meteorologists debated the phenomenon and its practical implications in professional meetings all over the world. Two years later, the UN group met again in Villach, and then in Bellagio, Italy.

In 1988 Canada took the initiative again, this time on the greenhouse effect, by inviting 340 participants from 46 countries to Toronto. Although sponsored by the Canadian government and attended by a hundred government officials, Canada billed the meeting as nongovernmental. The final conference statement declared, "Humanity is conducting an unintended, uncontrolled, globally pervasive experiment whose ultimate consequences could be second only to a global nuclear war...."[38] It went on to recommend (1) a 20 percent global reduction in carbon dioxide emissions by the year 2005, (2) a comprehensive international treaty, and (3) a fund paid for by the industrial countries to compensate for reductions in the third world. These three principles have been the center of the debate ever since. The industrial first world was willing to give the third-world countries privileged status on the rationale they had not benefited from the industrial revolution, and to a lesser degree, on the rationale that because they were not industrial, they did not contribute much carbon dioxide. This obviously ignored the huge amount emitted by China, second only to the United States. The Toronto Conference marked the transition of the global warming issue from a scientific forum to a governmental forum.

The White House, in organizing the American delegation, had learned its lesson from the Montreal negotiations and took the lead away from the State Department and EPA to manage the preparations directly. This time EPA was only one of several departments on a task force including the departments of energy, interior, and commerce, the OMB, and the Council of Economic Advisors, agencies that tended to favor business. The American delegation this time concentrated on the costs of reducing carbon dioxide emissions, unlike the Europeans who concentrated on the benefits to the environment.[39] Although the Toronto conference statement was decidedly green, it had no status as a treaty or governmental document. It was drafted by pro-environmental delegates and debated for less than one day.

Meanwhile the governments moved forward by establishing the Intergovernmental Panel on Climate Change (IPCC) under the auspices of the UN Environmental Program and the World Meteorological Organization. The United States strongly supported the IPCC. When the panel met in November, it organized three working groups on science, impacts, and response strategies. The response strategies group, which was chaired by an American, convened in Washington in January where President Bush's secretary of state, James Baker, addressed them, endorsing "prudent steps that are already justified on grounds other than climate change," soon known as the "no regrets policy." During his campaign, Bush had promised to be "an environmental president," and had attacked his Democratic opponent, Michael Dukakis, for not being green enough. In a television commercial filmed as Bush toured Boston Harbor by boat, he assailed Dukakis, the governor of Massachusetts, for not bothering to clean up the harbor even when grant money was available.

In North America, the summer of 1988 was the hottest in recorded history. Severe drought afflicted most of the Midwest and East. When the Senate Energy and Natural Resources Committee held hearings, James Hansen, who directed NASA's research team at the Goddard Institute, testified that "the greenhouse effect has now been detected and it is changing our climate now."[40]

GEORGE H. W. BUSH AND THE EARTH SUMMIT IN RIO

During the 1988 presidential campaign, Bush promised to be "an environmental president" and specifically declared his intention to counter the greenhouse effect with the "White House effect" and use the power of the presidency to overcome the challenge. His campaign director, James Baker, was personally committed to solving the problem. In the early months of the Bush administration, the environmental position seemed strong. Baker became the secretary of state, and in his first speech after taking office, said that preventing global warming would be one of his highest priorities, concluding, "Time will not make the problem go away."[41] Besides Baker, the new EPA administrator, William Reilly, pushed hard for an international agreement. The president's science advisor, Allan Bromley, joined those in favor when he was appointed a few months later. The Department of Energy was concerned about limitations on coal, but recognized the need for agreement. Yet after the first months of his term, Bush ignored the problem of greenhouse gases, and Baker backed away from his campaign enthusiasm.

Less than four months after the inauguration, the White House staff faced a problem when Hansen, the NASA scientist, was scheduled to testify on the effects of global warming before the Senate Subcommittee on Science, Technology, and Space, chaired by Democratic senator Al Gore. In his testimony the year before, Hansen had alarmed the senators about the dangers. Controversy erupted when Gore revealed that the White House Office of Management and Budget was censoring him this year. In his original text submitted for approval, Hansen had asserted that computer models showed that human activity would increase the temperature substantially, causing droughts and severe storms. The OMB, which routinely previewed executive branch testimony, directed Hansen to tone down his predictions by raising questions about its uncertainties and casting doubt on human activity as the cause. A later government witness, Dr. Jerry Mahlman, from the National Oceanic and Atmospheric Administration, also told the subcommittee that the OMB had tried to censor his testimony, but that he had resisted vigorously because the changes were "objectionable and unscientific." For the OMB to clear proposed testimony to a congressional committee is standard, and when news reporters inquired, the White House press spokesman claimed the decision was made "five levels down from the top."[42]

Within a few days, it became apparent that the "low-level" censor was none other than the president's chief of staff, John Sununu. Ironically, Sununu was one of the few scientists to reach a high post in government. For 16 years, he was a professor of mechanical engineering at Tufts University. Later he won election to be governor of New Hampshire. In 1988 his support helped Bush win the Republican primary, pivotal because it is the first of the campaign. Bush rewarded him for his early support and no-nonsense managerial skill by appointing him chief of staff. The debate over Hansen's testimony stimulated criticism that Bush was not furthering his campaign promises about global warming, and that the real policy of the administration was to do nothing. To parry this, Sununu had Bush instruct the American delegates to the IPCC, then meeting in Geneva, to invite experts from all over the world to a workshop in Washington. Although this brought favorable press at home, the European delegates in Geneva believed that Bush had sabotaged them, since they were ready to negotiate then, and the workshop would merely delay the process.

When the IPCC met in Washington in 1990, President Bush sprinkled his welcoming speech with caveats about economic growth, saying that "whenever possible, we believe that market mechanisms should be applied and that our policies must be consistent with economic growth and free market principles in all countries." He further offended environmentalists by commenting on the "uncertainty" of the scientific models.[43] The original draft of his speech had had a more environmental tone, however. The earlier version had the backing of Baker, William Reilly, and James Watkins, the secretary of energy, but it was opposed for being too hard on industry by Sununu, who rewrote Bush's speech.[44] News of his editing leaked to the press, and on Sunday he was questioned on television. Sununu claimed that he had intervened to counteract "a little tendency by some of the faceless bureaucrats on the environmental side to try and create a policy in this country that cuts off our use of coal, oil and natural gas."[45]

Two months later, Michael Boskin, head of the president's Council of Economic Advisors, weighed in against a treaty, predicting that it would cost the United States $200 billion a year and would batter economies worldwide. He went on to say that temperature increases of few degrees would not be important and would actually benefit agriculture.[46] The occasion was a two-day conference, of which Boskin was a co-chairman, of 18 nations to discuss the economic impacts of a treaty. The conference, which focused on costs, was outside the IPCC process. Environmental groups like the World Resources Institute and the Union of Concerned Scientists believed that the real aim of the Bush administration was to resist pressure for action. European countries proposed to stabilize carbon dioxide emissions at the 1990 levels. Some advocated that the industrial countries should pledge reductions before a treaty is signed, and to pay to help third-world countries to reduce their emissions. These goals matched the Toronto conference statement.

The meeting ended in controversy after the U.S. delegates handed out an unauthorized draft "Charter for Cooperation" that emphasized the scientific uncertainty of the warming. This outraged the Europeans. The American officials immediately withdrew the copies and claimed that it had been released by accident. To make the situation worse, someone leaked copies of confidential talking points the White House had given the top U.S. participants, which listed debating points like the uncertainty of the science and the huge expense and warned them to avoid arguing about how much warming was occurring.[47] Even without the U.S. blunders, the Europeans resented the American attempts to monopolize the conference and to block other viewpoints. The Bush administration's only support came from British Prime Minister Margaret Thatcher, but she soon broke ranks and proposed that the United Kingdom should reduce its carbon dioxide emissions enough to stabilize them at the 1990 level. Germany was proposing more, a reduction of 25 percent.

Using a strategy similar to the American conference designed to showcase its own perspective, the Europeans held meetings on global warming at the

Hague in 1989 and at Bergen, Norway, in 1990. EPA administrator Reilly attended at the Hague over the objections of Sununu. Americans were not invited to the Bergen conference. Although the Europeans at the Hague wanted to move forward on aid to the third-world countries, they pulled back because it would antagonize the United States. Over the following two years, more evidence accumulated supporting the greenhouse theory. The National Academy of Sciences reported that the threat warranted protective measures such as raising mileage standards for new cars, increasing energy efficiency of building and appliances, reforestation, and increasing nuclear electric generation.

Chief of Staff Sununu continued to oppose international cooperation, believing that the scientific evidence was flawed and its advocates were "hysterical." When a State Department official took the Bush and Baker campaign promises at face value, Sununu had him fired. His fierce opposition intimidated Reilly and Watkins. Baker, who faced many foreign policy problems of more immediate concern, decided not to waste his time and reputation on the issue. Sununu, true to his engineering instincts, had a model of the world climate installed on his personal computer. His objections to the warming predictions were that the science was inconclusive and that reducing emissions would harm economic growth. While serving as governor of New Hampshire, he had favored the economy over the environment. He believed that the advocates of reducing emissions were just the same old no-growth zealots now exploiting this fad.

Reflecting on his government service a few months after leaving the White House, Sununu explained that he had opposed the treaty because it was premature. He said again that the scientific proof was inconclusive, that carbon dioxide was not the most appropriate gas to target, that the potential of the ocean as a heat sink was ignored, and that cooling due to the explosion of the Mount Pinatubo volcano the previous year would delay warming (assuming it existed) by 20 years. Sununu claimed that a survey of scientists on the mailing lists of environmental groups found that only 13 percent believed in the need for any immediate action. He asserted that the real reason environmentalists wanted to reduce carbon dioxide was that it would reduce economic growth, and that they were trying to pressure political leaders by putting them on the spot at the 1992 Earth Summit. "I think the whole idea of the Rio conference was to corner the world leaders, especially the president of the United States, in an election year, and to force them to make a decision that, on a more rational basis, they might not take." Sununu condemned media coverage of science and technology issues as "pretty bad," observing that, "for the most part, articles have an apocalyptic tone that misses the mark in terms of science and technology." He said that while in the White House, "I argued for a billion-dollar research program," and resisted "this tremendous rush to have the world tie its hands today, instead of waiting for the results of that research." In response to a question on how his engineering training had helped in policy making, Sununu listed "quantitative intuition" and "an ability to break things down into their component parts, analyze them and put together a solution." "Too many people in politics," he concluded, "talk about where they want to go, without ever knowing where they are starting from."[48]

Attention turned toward the Earth Summit scheduled in June 1992 in Rio de Janeiro. The UN Conference on Environment and Development had invited 160 nations to attend. It was the successor to the first and only environmental summit held in Stockholm 20 years before. Presidents and prime ministers from every country were invited to Rio. To prepare, diplomats and scientists began meeting several months in advance at the UN headquarters in New York to draft a treaty limiting carbon dioxide emissions. Seventy-seven countries in the third world, known as the Group of 77, mobilized to block the first world from imposing carbon dioxide limits that would limit their economic growth and to seek money from the first world to pay for any sacrifices. Europeans and the Japanese agreed to reduce their carbon dioxide emissions to the 1990 levels by the year 2000. In addition to the problem of greenhouse gases, the New York negotiations addressed deforestation and endangered species.

In December Sununu had been forced to resign because of the embarrassment when the public learned he had misused the White House jet to visit his dentist in New York. At first Reilly, Watkins, and Bromley hoped that the policy would become more favorable toward a treaty, but pro-business forces were still powerful. To be the new chief of staff, Bush appointed Samuel Skinner, an accountant, prosecutor, and top IBM salesman who had been serving as secretary of transportation. In that job, Skinner had coordinated cleaning up Prince William Sound after the *Exxon Valdez* oil spill and favored tougher fuel efficiency standards for automobiles. He had a reputation for fairness and willingness to listen.

About the same time, Bush appointed Clayton Yeutter to head the White House Domestic Council, which advised him on policy. Yeutter had served in the Reagan administration as the president's special trade representative, where he promoted American exports, and more recently had been secretary of agriculture and chairman of the Republican National Committee. He had earned both a law degree and a PhD in agricultural economics and had had a career in international agricultural trade. Yeutter soon persuaded the president to refuse to visit Rio if the treaty had legally binding emissions limits affecting the United States. He announced that Bush would only attend the Earth Summit if were in the best interests of the United States. Yeutter was reinforced by Richard Darman, director of the Office of Management and Budget, who emphasized that carbon dioxide limits might harm the economy, at that point mired in a recession. Department of Energy computer models predicted a negative economic effect. When EPA pointed to the success of energy-saving programs, the White House interpreted this as showing less need for a treaty.

In preparing for the Rio Summit, the State Department was taking the lead, not EPA. Bowing to American pressure, the international negotiators in New York left their draft treaty ambiguous by establishing a framework for emissions limits without actually setting any numbers and without promising any money to the third world.[49] The treaty was named the Framework Convention on Climate Change (FCCC). In keeping with the custom for international law, the terminology "convention" refers to a multilateral agreement and framework refers to the expectation that an agreement will be an umbrella covering additional specific agreements, termed protocols, to be negotiated in the future.[50]

Foreign countries were quick to criticize President Bush. European countries and Japan decried this manipulation. Fernando Collor de Mello, president of Brazil and host of the Earth Summit, telephoned Bush to try to persuade him to change his mind. The chairman of the New York negotiations, the French diplomat Jean Ripert, diplomatically put the best face on backing down by saying, "We've got to start somewhere."[51] Most environmentalists considered the draft a sham. With the presidential election campaign underway, Governor Bill Clinton, by then the leading candidate for the Democratic nomination, attacked Bush for not backing a stronger treaty. Although the president was a friend of business and worried about the current recession, he also worried how the voters would take his antienvironmental stance. Besides, the globe-trotting president did not want to miss the largest gathering of world leaders in history, which 117 heads of state or government were scheduled to attend.

Addressing the Rio conference, an ebullient George Bush declared, "Twenty years ago some spoke of the limits to growth, and today we realize that growth is the engine of change and a friend of the environment.... America's record on the environment is second to none. So I did not come here to apologize. We come to press on with deliberate purpose and forceful action.... I'm happy to report that I've just signed that framework convention on climate change." In contrast, on the closing day Maurice Strong, the Canadian diplomat who organized the conference, said, "Our present economic system is not sustainable and I don't think that that message has got through to all leaders."[52]

During the fall campaign, Clinton continued to attack Bush's weakness on the environment. For his running mate, he chose Senator Albert Gore, widely known as a strong environmentalist. Yet once Clinton and Gore took office, the new administration did not move to change the Bush administration policies. With Earth Day approaching on April 20, Gore asked Clinton to promise in his speech that the United States would try to reduce its emissions to the 1990 levels by 2000, the very goal Bush had forced the UN negotiators to remove from its draft treaty for Rio less than a year before. EPA favored limiting emissions, but opposition came from the Treasury Department, worried about reducing economic growth, and from the Energy Department, concerned that it did not have enough data.[53] Clinton's compromise was to announce that the United States would undertake voluntary measures. The switch from a Republican to a Democratic administration had not ended the conflicts between business and environmental interests within the executive branch. The balance, however, tended more toward the environmental side.

THE KYOTO PROTOCOL

Notwithstanding Clinton's campaign rhetoric, his administration put global warming on the back burner for the next five years. From time to time the White House toyed with schemes such as a carbon tax to deter fossil fuel consumption, but never went further because it anticipated strong opposition from energy producers and consumers. Heavy industry, electric utilities, and labor unions argued that a carbon tax would cause production and jobs to move overseas to countries that did not restrict fossil fuels.

In 1995 the countries that had signed the FCCC in Rio held their first Conference of the Parties in Berlin. The Europeans wanted to push forward. Timothy Wirth, undersecretary of state for global affairs, told the conference that the United States also wanted to move forward. The conference issued the "Berlin Mandate" to begin a two-year analysis and assessment phase to negotiate a comprehensive menu of actions. One incentive for the Europeans was that a likely result of global warming would be disruption of the Gulf Stream, which brings warm water from the Caribbean to moderate the northern European climate. Temperature readings showed the polar regions would warm faster than the tropics and temperate zones, and one scenario was that higher temperatures would melt the sea ice on the Arctic

Ocean, turning the Gulf Stream north along the coast of Greenland instead of west toward the British Isles. Thus the ironic consequence of warming would be to lower the temperatures in northern Europe.

The following year the second Conference of the Parties continued its negotiations. The Europeans favored restricting the industrialized nations sooner by regulation beginning in 2005, and the United States favored waiting and using market methods like emissions trading. This would be similar to the sulfur allowances in the 1990 Clean Air Act Amendments, which permit companies to buy and sell the right to pollute a certain number of tons. The parties continued to debate the responsibilities of the third world. Again the underdeveloped countries argued that the first world has already enjoyed the benefits of 150 years of industrialization and that it was now their turn. Their greenhouse gases should not be restricted for another 50 or 100 years. In the end the conference adopted a ministerial declaration largely following the American position statement that (1) accepted outright the scientific conclusions of the IPCC, (2) rejected uniform standards in favor of flexibility, and (3) called for "legally binding mid-term targets." In other words, the conference declined to limit gas emissions until the year 2010 or 2020.

In anticipation of the third Conference of the Parties in Kyoto in December 1997, the European Union challenged the other industrial countries to reduce greenhouse gases by 15 percent below the 1990 level by 2010. Japan joined the United States to oppose this. The third-world countries continued to press for the first world to reduce emissions, but to exempt themselves. The Association of Small Island States, such as Samoa and the Maldive Islands, advocated this strongly, because they would be to first to suffer when the sea level rose. The highest point on the Maldives, an archipelago of 1,200 islands in the Indian Ocean, is six feet above sea level, and 80 percent of the tiny nation is less than three feet above sea level.

In the six months leading up to the Kyoto conference, President Clinton experimented with various stances. In June, feeling the pressure from the European Union, he addressed the United Nations, saying that the United States would back legally binding limits on emissions, although he was vague on the details. In July he backtracked after the Senate passed a resolution, 95–0, that it would not approve a treaty with emission limits. In October, to influence public opinion, Clinton invited 110 weather newscasters from the country's top 80 television markets to the White House to persuade them to publicize the danger of warming. NOAA officials briefed them, and the president and Vice President Gore gave pep talks. That evening the weather reporters broadcast live to their home stations from the South Lawn.

Clinton frequently has told how Al Gore converted him to the greenhouse theory. Gore wrote at length about climate change in his book, *Earth in the Balance.* The president's advisors, however, were split. Gore, Undersecretary of State Wirth, environmental advisor Katie McGinty, and EPA and Interior officials urged strong action. On the other side, the deputy secretary of the treasury, Lawrence Summers, and the chairman of the Council of Economic Advisors, Janet Yellen, opposed emission limits because they would dampen economic growth. In October Clinton announced that targets and time-tables for cutting greenhouse emissions should be postponed for 20 years.

By this point the climatological experts in the United States and else-where gave near unanimous support to the greenhouse theory, and only a few dissented. By 1997 the scientific evidence supporting the greenhouse effect was stronger than in 1988 or 1992. The computer models were better than ever. Meteorologists had learned more about the impact of water vapor, which both warms and cools. Since 1978, satellites anomalously had been recording cooling temperatures, not warming ones, but this was now explained as due to the cloud cover and inaccurate interpretation of the data. Re-analysis showed that two sudden dips were recorded at times when one satellite was replacing another.[54] Other studies had weakened alternative theories, such as the sun's cycles of activity or the orbit of the earth. The world scientific community lined up overwhelmingly on the side of the greenhouse theory.

The debate within the American government continued until the delegation departed for Kyoto, and then a bit longer. The strategy was to hold firm against European demands for a legally binding timetable and to insist on a market in emissions. At first Gore was not scheduled to be part of the delegation because his well-known pro-environmental stance might alarm business interests, but when the negotiations bogged down after a week, Gore persuaded the president to let him go, after promising he would not become personally involved in the negotiations. Once he arrived, however, he instructed the U.S. delegation to exercise "increased negotiating flexibility." It was hard for the government officials in the room to resist the vice president's hints. Gore himself had a lot at stake politically. His environmentalism had won him national visibility and, some claim, the vice presidential nomination. His ambition to run for president in 2000 was well known. The vice president's environmental supporters were disappointed with his waffling, viewing his trip as opportunism and publicity grabbing, and on the other side, business and union leaders believed he had failed to protect jobs and production.

Under the final Kyoto Accord on Climate Change, most industrialized countries agreed to reduce greenhouse gas emissions 6–8 percent below the 1990 levels by 2008–2012. Developing countries had no restrictions until at least 2005. Although the following year the fourth Conference of the Parties was scheduled to discuss the issue in Buenos Aires, little came of this session. The Kyoto Accord provided that the industrial countries would be able to participate in emissions trading once a market was established. Debate on this was intense, with India and China arguing that trading would allow industrial countries to escape making the required cuts. India, China, Mexico, and Brazil managed to block a U.S. proposal to make them subject to restrictions because they emit so much carbon dioxide. In spite of its third-world status, China releases more than any country except the United States. The first world agreed to set up a "clean development fund" to give money to the third world.

Once the conference ended, Clinton was in no hurry to send the treaty to the Senate for ratification, especially because of the 95–0 vote against it the previous July. The president suggested that a one-year wait seemed appropriate. His rationale was that by then the fourth Conference of the Parties meeting in Buenos Aires would resolve the issue of the responsibilities of the third world, but in fact this did not occur, and the real reason for the delay was a lack of votes. When Clinton left office three years later, he had still not submitted the treaty to the Senate.

This was not so much lack of will as paralysis. After realizing that the Kyoto Protocol was doomed, Clinton and Gore toyed with the idea nicknamed "cap and trade." Governments would sell carbon emissions permits in an unlimited number, but with a cap on the price per ton, which would keep costs down and compliance up. The administration informally sounded out this concept on the Europeans, who gave it some support. The barrier was American environmental groups, who were staunchly opposed. Consequently, Gore decided to do nothing. The election was coming nearer and he did not want to have a controversy with one of his core support groups.[55] Keeping the lid on this particular environmental dispute was not enough to give him victory in the extremely close electoral vote against George W. Bush. Ralph Nader, the longtime consumer crusader and gadfly, had gained the presidential nomination from the new Green Party and won nearly 3 percent of the vote. This was enough to tip a few states from the Democratic to the Republican column, for example Florida, where Nader received more than 90,000 votes. This occurred even though the mainline environmental groups like the Sierra Club had anticipated the danger and had endorsed Gore.

Once in office, George W. Bush did not waste much time before renouncing the Kyoto Protocol. On March 13 he rather defiantly attacked the treaty, and his National Security Council director, Condoleezza Rice, announced that it was "dead." His rejection fit his emphasis on burning coal as part of his energy plan, a central planning concept. In the preceding months California had suffered electricity blackouts due to lack of generation and transmission capacity. Later evidence pointed to price manipulation as well. Bush had just charged Dick Cheney with preparing an energy plan, which turned out to propose increasing coal and oil supplies. One of its options was to drill for oil in the Arctic National Wildlife Refuge in Alaska, long a red flag for environmentalists. They also criticized his attack on the Kyoto Protocol for being "anti-science." The president appeared to be ignoring other instances of scientific advice, such as by restarting a largely abandoned Reagan Administration Strategic Defense Initiative, labeled Star Wars, that had proved to be a technological fantasy.[56]

Bush wrote, "I oppose the Kyoto Protocol because it exempts...China and India...and would cause serious harm to the US economy."[57] Strong economic growth in the prior eight years had increased the American emissions of carbon by 30 percent. In other words, industry and transportation would have to decrease fuel consumption by 37 percent to reach the Kyoto level (a 7 percent reduction below 1990). Even if the entire world met the Kyoto goals, total emissions would increase 26 percent (compared to an uncontrolled 34 percent) according to an earlier study by the Energy Information Administration. The cost of electricity would rise 86 percent, and the cost of gasoline would rise 53 percent.[58]

The president said he could not act on global warming "given the incomplete state of scientific knowledge of the causes"[59] and asked the National Academy of Sciences to re-analyze the information, a tactic many considered phony. The IPCC had just reported a few months earlier that "there is new and stronger evidence that most of the warming observed over the past 50 years is attributable to human activities."[60] The academy did re-analyze and got its report back in record time, saying that carbon dioxide and other greenhouse gases could heat the planet between 3 and 10 degrees Celsius, and that the shift "could well have serious adverse societal and ecological impacts."[61] The president's response was more study. He proposed a national climate change technology initiative through the Energy and Commerce departments to conduct basic research at universities and the DOE laboratories. The Europeans were quite angry at Bush's rejection of the protocol.

It took a while for EPA to fully get the message that the Bush administration totally opposed the Kyoto Protocol, or even the issue of warming. In June 2003 a controversy emerged between EPA and the White House over the publication of a summary report on environmental conditions in general. Two years before, the administrator, Christine Todd Whitman, had commissioned the comprehensive report to cover air, water, land, human health, and ecological interactions. Although Whitman was a loyal Republican, she had become frustrated about the failure of the administration to protect the environment. In her two years as EPA head, she had continually lost out to business and industrial forces. So she announced her resignation to take effect on June 27. This meant that the report would be her valediction.

In late April, EPA sent a draft to the White House with a short section about how greenhouse gases were causing global warming. The White House edited out these objectionable portions and proposed replacing them with a few paragraphs that came to no conclusion. The editing eliminated references to many studies concluding that warming was due to carbon emissions and could threaten health and ecosystems. The editing removed conclusions of the 2001 report on climate by the National Academy of Sciences that President Bush had requested and that he had endorsed in speeches that year. The White House also deleted a reference to a 1999 study showing that global temperatures had risen sharply in the previous decade compared with the last thousand years. In its place, White House officials added a reference to a new study, partly financed by the American Petroleum Institute, questioning that conclusion. Whitman decided it would be better to have no climate change section at all than the one proposed by the White House, which she characterized in an interview with the *Los Angeles Times* as "pablum." An internal EPA memorandum that was leaked to a newspaper said that the White House version "no longer accurately represents scientific consensus on climate change."[62] When the report was released to the public on June 30 the warming problem had disappeared. The section on global issues discussed the ozone hole and nothing else. The companion "technical document" had a single sentence about climate change that said, "This report does not attempt to address the complexities of the issue."[63]

With the United States rejecting the Kyoto Protocol, the Bush administration hoped that the protocol would never go into effect. To do so it required ratification by countries that emitted 55 percent of the world total. The last hope to prevent this was Russia, with 17 percent of the world emissions of greenhouse gases. Its approval would bring the total emissions for the signing parties to over 55 percent. After playing coy for several years, Russian president Putin announced that his country would ratify the Kyoto Protocol in return for European support for its admission to the World Trade Organization. "The fact that the European Union has met us halfway at the negotiations on membership in the WTO cannot but influence Moscow's positive attitude toward ratification of the Kyoto protocol."[64] Half of Russian external trade is with Europe, and a fifth of European external trade is with Russia. The expansion of the union to 25 members brought its boundaries right to Russia's western border. Furthermore Russia stood to gain by selling its emissions in excess of its target to other industrial countries. Because the quotas were based on 1990 levels, it was unlikely that Russia could ever get such a good deal again. Thus, with Russia putting the voting over the top, the protocol officially went into effect in 2005.

President Bush was disappointed that the protocol went into effect. For four years Russian reluctance had meshed with the administration policy. Now the only industrial country to oppose the protocol was Australia, with emissions of only 330,000 tons per years, the same as South Africa and Brazil. Prime Minister John Howard seemed to enjoy being Bush's partner in opposition. Bush and Howard cooperated in creating a group of Pacific nations that favored an alternative to Kyoto, called the Asia–Pacific Partnership on Clean Development and Climate. The others were Japan, China, India, and South Korea. The partnership proposed to use more renewable energy and improve efficiency as its contribution to reduce global warming. Although Japan had signed the protocol as an Annex I (that is, industrial) participant, it did not seem to see joining the partnership as a contradiction. The other four were not on Annex I, that is, they were considered underdeveloped, hence exempt. China, of course, emits 3.5 million tons of carbon dioxide a year.

More bad news for the protocol came from Britain. The government had strongly supported controlling greenhouse gases since the early 1980s when Margaret Thatcher was prime minister. Although she was a Conservative and a leader in promoting business, she was also a chemist by training, who early recognized the danger. Support continued when the Labour Party won election in 1997 and Tony Blair became prime minister. Indeed his advocacy was even stronger. Yet a strange thing happened with Blair: He lost his enthusiasm for the Kyoto Protocol. In 2005 he said that he was "changing my thinking about this." He no longer believed that negotiating international treaties was going to help. "The truth is," he observed, "no country is going to cut its growth or consumption substantially in light of a long-term environmental problem."[65] The only hope he saw was new science and technology. He said:

We also have to recognise that while the Kyoto Protocol takes us in the right direction, it is not enough. We need to cut greenhouse gas emissions radically but Kyoto doesn't even stabilise them. It won't work as intended, either, unless the US is part of it.... We have to understand as well that, even if the U.S. did sign up to Kyoto, it wouldn't affect the huge growth in energy consumption we will see in India and China.[66]

Back in the United States, Hurricane Katrina's devastation of New Orleans sparked debate on the role of global warming on hurricane frequency and intensity. These storms are generated when the temperature of the tropical Atlantic is above 78 degrees F. That year the Gulf of Mexico was two degrees above average, and Katrina picked up energy after it crossed Florida and swung through the gulf. The day before it hit Louisiana, it reached category five, the very highest. Two other hurricanes, Rita and Wilma, also reached category five. The season had 28 tropical storms, so many that the scientists ran out of names and had to name the last few for Greek letters. Yet the number in the previous decade had not been greater than average, and the following season had few. The other aspect of hurricanes is their intensity, and here the evidence is stronger. Recent ones have been more powerful. Analysis by the National Center for Atmospheric Research concluded that warmer ocean temperatures were the cause, not natural cycles.

Katrina added fuel to the controversy over the "hockey stick." This refers to a dramatic graph that gained prominence when published in the 2001 *Third Annual Assessment* of the IPCC showing a long-term gradual increase in global temperatures, but with an abrupt turn upward in recent years. The original graph was calculated by a team of climatologists led by Michael Mann. It displays temperature data going back 2,000 years. The controversy is not about the long-term gradual increase, but the rapid increase recently. Critics claim the evidence is too scanty. In particular they point to the absence of the medieval warming and the little ice age in the sequence. But Mann counters that those were regional, not global, phenomena. The scrappy climatologist says, "From an intellectual point of view, these contrarians are pathetic, because there's no scientific validity to their arguments whatsoever."[67]

Mann had been in the news in 2003 at Senate hearings. The topic was a bill to require U.S. industry to reduce its emissions of carbon dioxide and five other greenhouse gases sponsored by John McCain, a Republican, and Joseph Lieberman, a Democrat. Neither senator was a friend of President Bush. McCain had opposed him in the primaries in 2000, and Lieberman had been the Democratic nominee for vice president. McCain had never paid any attention to the global warming issue until he ran for president. As he campaigned at rallies for the Republican nomination, he kept encountering the Polar Bear, a heckler dressed in costume who wanted to publicize the problem. At the hearings, the committee chair was Senator James Inhofe, a Republican who was loyal to Bush, and he denied the scientific basis of global warming. Inhofe represented Oklahoma, a state with a large oil industry.

One witness called before the committee was Willie Soon of the Harvard Smithsonian Center for Astrophysics, who had recently authored an article reviewing 240 articles on global warming, coming to the conclusion that the warming of the twentieth century was not unusual in comparison to the past thousand years, and that the cause was not human activity. Inhofe enthusiastically endorsed Soon's argument and rejected that of Mann, another witness, who testified that mainstream climate researchers had concluded that the recent warming was unprecedented and was caused by human activity. Inhofe described those scientists as "the Chicken Little crowd" and told the entire Senate that "with all of the hysteria, all of the fear, all of the phony science, could it be that man-made global warming is the greatest hoax ever perpetrated on the American people? It sure sounds like it."[68] Although McCain was the co-sponsor of this particular bill, virtually all other Republicans lined up behind President Bush, producing what some called Republican and Democratic versions of science.

Surprising support for controlling global warming came at the state and local levels, even while the national level rejected Kyoto. California enacted a law to reduce emissions by 25 percent by the year 2020. The Air Resources Board is to require mandatory reporting and design a cap and trade program to go into effect in 2012. Chicago, Los Angeles, Philadelphia, and New York pledged to reduce their emissions. The U.S. Conference of Mayors passed a resolution to reduce sprawl and improve efficiency. Whether these extend beyond pious hopes is unclear.

CONTRARY VIEWPOINTS

At the time of the Kyoto Conference the theory of the greenhouse effect had so much scientific consensus that most observers were startled when the *Wall Street Journal* published an opinion piece a few days before the meeting announcing, "Science Has Spoken: Global Warming is a Myth."[69] The conclusion fit the *Wall Street Journal*'s pro-business editorial policy. The authors were Arthur and Zachary Robinson, identified as chemists at the Oregon Institute of Science. At the same time, thousands of scientists and officials received a copy of "Environmental Effects of Increased Atmospheric Carbon Dioxide" by the same authors. The cover letter, signed by Frederick Seitz, Past President, National Academy of Sciences, and President Emeritus of Rockefeller University, warned against the draft treaty and asked the recipient to sign a postcard opposing the treaty. In fact, the "article" was fake. Although printed to resemble an article published by the National Academy of Sciences, it was phony. It was privately printed, was not endorsed, and was not reviewed by peer scientists.

The Robinsons argued along five lines. First was that the earth was not warming, pointing out that temperatures measured in the troposphere are declining and that surface temperature readings are often made in cities that are "heat islands" due to buildings and pavement. They claimed that any warming in the past century was only part of an increase since the little ice age three hundred years ago and was far below the temperatures during the medieval climate optimum nine hundred years ago. They accused alarmists of selecting partial data, for example by neglecting data from the southern hemisphere, which has not warmed, or using selected years and not the complete record. Second, the Robinsons argued that most of the temperature rise in the twentieth century occurred before 1940, whereas most of the carbon dioxide rise occurred after 1940. Third, they debunked claims that global warming is causing more hurricanes and storms or would increase sea level. They gave evidence that the number of hurricanes had decreased since 1940 and that the maximum wind speeds had remained the same. They maintained that historical records failed to show any rise in the sea level, and that the fear that the Antarctic ice cap would melt is not consistent with experiments or theory. The Robinsons argued, fourthly, that increased carbon dioxide was good because it would increase the growth of plants. Experiments demonstrated that extra carbon dioxide enhances the growth of pine trees and orange trees. Plants can grow better in arid regions. The Robinsons' concluding argument was the importance of fossil fuels in increasing wealth and prosperity everywhere. People with more income are better nourished, healthier, longer lived, better educated, and more productive. Moreover, these people will live in a world more lush with plant and animal life.

Aside from Willie Soon and the Robinsons, only a handful of scientists maintain that the greenhouse theory is wrong. Richard Lindzen of MIT said that the evidence of warming was inadequate.[70] Klaus Hasselmann, director of the Max Planck Institute for Meteorology in Hamburg, Germany, maintained, "We simply don't have sufficiently reliable estimates of the natural variability" to determine the connection between greenhouse gases and the warming trend.[71] Bjorn Lomborg contended that the evidence was inconclusive, then made his main argument that even if it were a problem, remediating it would be a waste of money, and that a better course would be to solve present-day problems of poverty and health for the third world. Knud Lassen and Eigil Friis-Christensen of the Danish Meteorological Institute argued that the solar cycle, not human activity, causes temperature variations.[72] Others assert that ocean currents are the key.[73]

Patrick Michaels of the University of Virginia and the Cato Institute is perhaps the most articulate skeptic. He maintained that the historic record was distorted and that most of the increase in temperature came before 1945, whereas most of the increase in carbon dioxide came after 1945, a period in which the temperature was stable.[74] He said that cloud cover plays a bigger role than "mainstream scientists" believe, and that the actual temperatures do not agree with those predicted by the computer models. The actual increase in the past hundred years has been only one degree F, only a third of the three to four degrees predicted. Michaels asserted that the computer models were "'adjusted' arbitrarily in order to keep them from producing unrealistic climates." Generally this involves adding an increment of heat moving from the tropics to the poles. Michaels claimed that even if one accepted the theory, "The amount of warming 'saved' by the emissions reductions mandated by the Kyoto Protocol is infinitesimal."[75] Moreover, he pointed out that the exemption for China, India, and the Pacific Rim omits 25 percent of the carbon dioxide emissions. Overall the minute reductions of the Kyoto Protocol pale in comparison to the costs. The effort will decrease economic productivity by 12 percent per year.

The Cato Institute, where Michaels is based, is a conservative think tank, dedicated to finding solutions to policy problems "that are consistent with the traditional American principles of limited government, individual liberty and peace." It favors using the market and describes itself as Libertarian. The institute, which flourished during the Reagan administration and again during the Bush (Junior) administration, considered the Clinton administration too prone to "statist" solutions. It has enjoyed generous support from conservative business interests.

The Heartland Institute, based in Chicago, has a similar orientation, although is lacks the prestige of the Cato Institute. It argues that the data does not show a warming trend, and that even if it did, human activity is not proven to be the cause. Furthermore, a modest increase in temperature would be beneficial to agriculture, industry, and life style. The sea level would not rise due to greater evaporation. Another opponent of controlling global warming is the George C. Marshall Institute. Its chief executive officer is William O'Keefe, formerly the executive vice president of the American Petroleum Institute, and formerly head of the Global Climate Coalition. The Marshall Institute believes that a strong economy is more important than stopping the warming.

Going beyond technical arguments to political action, the huge Global Climate Coalition mounted a major lobbying effort from 1989 until 2002. Direct membership consisted of 60 associations and businesses. Associations included the Edison Electric Institute, the American Petroleum Institute, the U.S. Chamber of Commerce, and the National Association of Manufacturers. Corporate members included the Big Three auto manufacturers, big electric utilities, and oil companies like Exxon and Amoco. Through its association members, the coalition claimed to represent 230,000 businesses. It maintained a professional staff in Washington that monitored Congress, the White House, and the executive branch agencies. The coalition sent a large staff to the Kyoto conference to lobby the delegates and coordinate those opposed to the protocol. The organization's position proved so strident that major members disaffiliated, including American Electric Power, BP, Shell, and the Big Three auto manufacturers. They were getting too much bad publicity. The defections, combined with the election of George W. Bush and the inability of environmentalists to persuade Congress to act, contributed to its decision to close.

The Edison Electric Institute objects vigorously to the Kyoto Protocol. It is the trade association for privately owned electric-generating utilities. Because coal produces over half of the nation's power, the industry has a lot to lose if it is restricted. The EEI objects to the protocol because it exempts the underdeveloped countries and does not even have a process for the future for moving them to binding limits or give credit for their voluntary reductions. Furthermore it will be hard on the U.S. economy, excludes forest management as a technique, is too vague about emissions trading, and has no place for voluntary action. The institute estimates the treaty could cost the American economy $100–200 billion per year by 2010. Instead, the EEI recommends voluntary programs such as renewable energy, geothermal geysers, demand management, and forestry. Considering that nuclear power plants emit absolutely no carbon dioxide, the institute is strangely muted. In France, 77 percent of the electricity is nuclear, in contrast to only 20 percent in the United States.

A number of electric utility leaders believed that the EEI did not push hard enough, so in 2001 they organized the Electric Reliability Coordinating Council. Members included American Electric Power, the Southern Company, TVA, TXU from Texas, First Energy, Duke Energy, and Progress Energy. The leader was Haley Barbour, at the time a lobbyist representing several of these corporations, and formerly the chairman of the Republican National Party. Individuals and political action committees from five of these companies donated more than $1 million each to political parties, favoring Republicans with nearly three-quarters of their money. Energy company executives have been particularly enthusiastic about the Bush candidacy for president. They constitute the largest group within the Pioneers, the name given to elite Republican fund raisers who each generated more than $100,000 in donations. When in 2001 the energy companies became concerned that EPA was not sufficiently sympathetic, Barbour wrote to Vice President Cheney to complain.

The petroleum industry has nearly as much to lose as electric utilities. Less than two months after the Kyoto conference, the American Petroleum Institute launched a public relations campaign to spend millions of dollars to convince Americans that the draft protocol was based on shaky science. One tactic was to recruit a cadre of scientists who shared the industry's view and train them in public relations so they could help convince journalists, politicians, and the public that the risk was too uncertain and distant to justify controls on greenhouse gas emissions. On the other hand, in the wake of Kyoto, leaders of major companies have begun to acknowledge the problem. John Browne, CEO of British Petroleum, said, "We've moved (as the psychologists would say) beyond denial," and reported that senior people in the industry like the heads of Royal Dutch Shell and Texaco were now looking at the consequences and options.[76] BP has incorporated statements in support of antiwarming policies in its advertisements.

While those who reject the warming theory totally clearly go against the mainstream of the present scientific consensus, at the other extreme are environmentalists and climatologists who believe that the warming will be sudden and dramatic. They are alarmed about weather anomalies like Katrina, El Niño, droughts, floods, and hurricanes. They worry that the present climate is not robust, but delicately balanced so that a slight warming could throw off the entire system. Instead of gradual warming that will takes at least a century, the climate will fluctuate wildly, with storms sweeping the earth, the Greenland ice cap melting quickly, and the Gulf Stream no longer carrying Caribbean warmth to Europe. At present, this extreme view is not very noticeable because its proponents are lumped with the mainstream scientists. Indeed many environmentalists consider the alarmist position beneficial because it strengthens the case for controlling greenhouse gases.

The problem with the alarmism is that there is little evidence to support it. The common threats like storms and droughts are not any more common than in the past. Of course, accurate data on past weather is not dependable. Although meteorologists may have good records for Europe and much of North America for the past century or more, they do not know much about the history of the rest of the world. Improved technology for measuring ocean currents and taking readings from satellites is only two decades old. Moreover, many of the weather disasters the alarmists cite are the same ones used to argue the case for global cooling in the 1970s.

One organization has staked out a middle ground of good science and moderate policy. The Pew Center for Global Climate Change, organized in 1998, has membership of 38 major corporations, a number of which had left the Global Climate Coalition. These include Royal Dutch Shell and British Petroleum. Three major electric-generating utilities joined, and other members included Alcoa, Boeing, and Georgia-Pacific. The American auto manufacturers did not join, but Toyota did. The Pew Center acknowledged that human activity was the cause of warming. The center notes that it is nonpartisan and independent. It was organized by the Pew Charitable Trusts, which originally got its money from the Pew family, owners of the Sunoco oil company. Today the center gets funds from a variety of charities and individuals, but does not accept corporate donations.

The Pew Center seeks to reach out to progressive industry and to influence policy. Its techniques are policy analyses, briefings, and press releases. It does not lobby Congress. The president is Eileen Claussen, who served as assistant secretary of state for oceans and international environmental and scientific affairs in the Clinton administration, and before that served on the staff of the Clinton White House. The center has a staff of 20, making it the largest concentration of climate change policy specialists outside the government. It does not see itself as an environmental group, but as halfway between them and industry. It maintains that the Kyoto agreement represents a first step, but more must be done both to implement the market-based mechanisms and to more fully involve the rest of the world. Recent reports have looked at the period beyond 2012. Some have described the Pew Center as doing the work EPA ought to be doing if it were not following the Bush administration policy.

THE EUROPEAN PERSPECTIVE

The European perspective differs from the American due to geography, trade, and attitudes toward market solutions. The countries in the northern portion of the continent depend on the moderating effect of the Gulf Stream, which is threatened by global warming. London, Amsterdam, and Berlin lie as far north as Calgary, Alberta; and Oslo lies as far north as White Horse, Yukon. Without the Gulf Stream, they would have frigid climates. In coming years, as the European Union integrates even more economically, members will have to compete more directly against each other, and against the United States. Therefore the European Union wants the other industrial countries to face equivalent burdens. It is less concerned about the third world because it does not compete directly. Moreover, the Europeans are more accustomed to government playing a major role in the economy and collecting more taxes. A carbon tax would enrich the government coffers at the same time as it discouraged emissions.

In a round of pledges in the mid-1990s, Germany volunteered to reduce emissions the most: 25 percent. While its commitment appeared noble, several factors made this less of a sacrifice than it appeared at first. Its reunification with East Germany in 1990 threw its economy into recession, reducing manufacturing and therefore reducing its emissions, for which it receives credit. In the east the antiquated heavy industry was shut down because of its low productivity. The Communist factories were inefficient and depended heavily on coal. In both east and west Germany, coal mines are nearly depleted. In response to the guilt remaining from World War II, Germany likes to be seen as a leader on moral issues. It has the strongest Green Party in the world. Most of its scientists work for the government (unlike the United States) and respect the wishes of their employer. Chancellor Helmet Kohl had a strong personal commitment to fighting global warming. A year after the Kyoto negotiations, Kohl's conservative party lost the general election to the Social Democrats. To gain a majority in the Parliament, the Social Democrats formed a coalition with the Greens, thus solidifying support for the protocol.

Britain promised to reduce greenhouse gas emissions by 20 percent, the second highest. Former prime minister Margaret Thatcher was an early convert to solving the problem, promising a reduction of 10 percent, and Prime Minister Tony Blair doubled the amount when the Labour Party won the 1997 election. The British economy is switching to natural gas from the North Sea (which produces less carbon) and, like Germany, faces depletion of its coal mines, which were heavily subsidized. British scientists have been on the forefront of research on global warming, and an Englishman, John Houghton, chaired the IPCC scientific panel for many years. France is not required to reduce emissions on the rationale that its extensive use of nuclear power has already contributed enough. The Dutch fear that warmer temperatures will raise the sea level and flood their country and have already increased the height requirement for their dikes. The policymaking systems of the European countries are not as open to public scrutiny as the American system. Industry is brought into the government discussions at an early stage, the number of scientists is smaller, they are less likely to raise objections publicly, and the European parliaments do not have the American tradition of free-wheeling investigations and exposés.

In negotiating at Kyoto, the Americans insisted on tradable emissions permits. These could be bought, sold, and traded between companies and even between countries. The Europeans are more accustomed to direct government regulation. Europeans and the Japanese considered the carbon trading unnecessary or worse. They believed that it did not reduce emissions and allowed the Americans to buy their way out of pollution. They believed that without trading, many old factories would close anyway, thereby ending their emissions, but that with trading, other companies would buy the rights and so perpetuate the emissions. Although the parties at Kyoto agreed to trade emissions, the exact details were not spelled out. Under Article 3, emission trading is only between the industrial countries. After all, the underdeveloped countries have no obligation to control emissions so they have nothing to trade. As supposedly industrial countries, Russia and Ukraine received allotments, which proved to be overly generous due to their economic decline. First of all, this made them look good in terms of compliance, and second, it gave them the opportunity to trade. The generous allowances for these countries gained the nickname of "hot air."

European countries, especially Germany and France, do not share the American enthusiasm for market techniques in the protocol. Besides trading, two others are clean development mechanisms and joint implementation. The clean development mechanism provides for an industrial country (Annex I) to form a partnership with a developing one to build a clean new factory but have the credit for less carbon go to the Annex I country. Joint implementation is a similar partnership, but between two industrial (Annex I) countries. The Europeans object both that it is immoral and that new factories are always cleaner than old ones, so no real benefit occurs. The protocol also provided for reducing carbon by storing it in sinks like forests or the soil. The Europeans do not have much room to increase their forests. Moreover they see cheating and tricky accounting to be likely. All of these techniques need detailed enforcement by scientists and greenhouse gas policemen. Strong enforcement would practically require world government. If the nations of the world could truly establish a world government, one would hope its first job would be to eliminate war.

By the time of the 2006 Conference of the Parties in Nairobi, talk moved to the period beyond the end of the first commitment period in 2012. New targets were supposed to be agreed to for the years following. In fact, virtually none of the industrial countries were meeting the target they had already pledged. With economic growth, Britain, Germany, and the rest of the European Union were hopelessly behind. Many predicted that there would not be any Kyoto Protocol at all. The parties held meetings at which the developing countries exhorted the industrial ones to do more. The Group of 77 said Annex I commitments for the second period should be substantially stricter. China said the second period should be longer. Brazil said that negotiations should simply lead to deeper commitments by Annex I parties and should not reopen previous agreements. All of them opposed having the non-Annex I parties take on commitments.

CONCLUSION

The heart of an apocalypse is a prophecy, and few can top Arrhenius's prediction that "we are evaporating our coal mines into the air." His forecast that the atmosphere would heat up by several degrees was marvelously prescient. Although Arrhenius viewed the warming with equanimity, most later scientists sounded the alarm. The resolution of the 1988 Toronto Conference described the danger of greenhouse gases as "second only to a global nuclear war." Al Gore wrote that the future of the earth hung in the balance. Many apocalyptic threats are recognized in advance by government commissions and reports. In the recent period, the Wallach Workshop of 1985 marks the time at which a scientific consensus arose. Following that one or two more technical conferences led quickly to the Toronto Conference and the IPCC.

The plan of how to deal with global warming sprang full blown from the Toronto Conference and has been little changed since. It was to limit emissions, exempt the third world, and give money to the third world. Eight years later it was enshrined in the Kyoto Protocol. This was a comprehensive plan for three-quarters of the earth. But it did not cover China, India, and the underdeveloped world, and hence left out a big and growing segment. Although the protocol officially went into effect in 2005, this was an empty victory. The United States, responsible for a quarter of the global emissions, did not participate. Even the countries that signed did not do much to limit their emissions, and most of them were emitting considerably more than when they signed in Kyoto and were not on track to meet their quotas for the end of the first commitment period in 2008–12.

From the Toronto Conference to the Kyoto Protocol, the potential greenhouse solution had the excellent model of the ozone solution. Both were atmospheric problems, both depended on scientific data, and both needed nearly complete international cooperation. Ozone was simpler, however. The time from discovery to resolution was faster—only a few years—and the scientific data was clear. Global temperatures, in contrast, warmed until about 1940, then cooled for a few years, then warmed and cooled, until about 1977 when the warming resumed and scientists could measure a clear phenomenon. Half of all CFCs were produced in the United States, in contrast to only a quarter of the greenhouse gases. This meant that an American solution was half of the world solution. Another way ozone was simpler was that industry soon invented alternatives to CFCs that proved to be cheaper.

NOTES

1. Quoted in Andrew Revkin, *Global Warming* (New York: Abbeville Press, 1992), p. 57.
2. Donella H. Meadows et al., *The Limits to Growth* (New York: Universe Books, 1972), pp. 71–73.
3. Elisabeth T. Crawford, *Arrhenius: from Ionic Theory to the Greenhouse Effect* (Canton, MA; Science History Publications, 1996), pp. 148–155.
4. Ibid., p. 155.
5. Revkin, *Global Warming*, p. 90.
6. Roger Revelle and Hans Suess, "Carbon Dioxide Exchange between Atmosphere and Ocean and the Question of an Increase of Atmospheric CO2 During the Past Decades," *Tellus* 9 (1957): 18-27.
7. Revkin, *Global Warming*.
8. Elizabeth Levy, *The SST Story* (New York: Delacorte Press, 1973), pp. 7–9; Mel Horwitch, *Clipped Wings: the American SST Conflict* (Cambridge, MA: MIT Press, 1982), pp. 27.
9. Ibid., pp. 220–224.
10. Levy, *The SST Story*, pp. 29–33.
11. Ibid., p. 93.
12. Ibid., p. 96.
13. Richard Elliot Benedick, *Ozone Diplomacy*, 1st ed. (Cambridge, MA: Harvard University Press, 1991), p. 21.
14. Quoted in Lydia Dotto and Harold Schiff, *The Ozone War* (Garden City, NY: Doubleday, 1978), p. 152.
15. Arjun Makhijani and Kevin R. Gurney, *Mending the Ozone Hole* (Cambridge, MA: MIT Press, 1995), p. 271.
16. Dotto and Schiff, *The Ozone War*, pp. 231, 239, 241.
17. Ibid., p. 303.
18. "Vienna Convention for the Protection of the Ozone Layer March 1986" (New York: UN Environmental Program, 1986), p. xxx.
19. Benedick, *Ozone Diplomacy*, pp. 82–83, 93; Montreal Protocol, Articles 2 and 5, www.unep.ch/ozone/pdf/Montreal-Protocol2000.pdf.
20. Benedick, *Ozone Diplomacy*, p. 96.

21. Ibid., pp. 46, 53.

22. Joseph C. Ferman et al., "Large Losses of Total Ozone in Antarctica," *Nature,* May 16, 1985, pp. 207–210.

23. *Montreal Protocol,* Article 10 as amended.

24. *Congressional Record,* April 13, 1994, pp. E639–640.

25. R. Monastersky, "Ozone on Trial," *Science News* 148 (October 1995).

26. G. Taubes, "The Ozone Backlash," *Science* 260 (June 11, 1993), p. 1580; Rush H. Limbaugh, *The Way Things Ought to Be* (New York: Pocket Books, 1992), p. 154.

27. U.S. House of Representatives, Committee on Science, Subcommittee on Energy and Environment, "Stratospheric Ozone: Myths and Realities," *Hearing on Stratospheric Ozone,* testimony by S. Fred Singer, September 20, 1995.

28. Ibid.

29. Ibid.

30. National Oceanic and Atmospheric Administration, National Climate Data Center, Global *Temperature Index* (www.ncdc.noaa.gov/ol/climate).

31. Central Intelligence Agency, Office of Political Research, "Potential Implications of Trends in World Population, Food Production and Climate," cited in House of Representatives, Subcommittee on International Organizations of the Committee on International Relations, *Prohibition of Weather Modification,* 94th Cong. July 29, 1975, p. 39.

32. Cf. Jean M. Grove, *The Little Ice Age* (London: Routledge, 1988), pp. 1–5, 376, 413.

33. Lowell Ponte, *The Cooling* (Englewood Cliffs, NJ: Prentice Hall, 1976), pp. 3–4; Paul Andrew Mayewski and Frank White, *The Ice Chronicles* (Hanover, NH: University Press of New England, 2002), pp. 131–133.

34. National Research Council of the National Academy of Sciences, *Changing Climate,* October 1983.

35. U.S. Environmental Protection Agency, Office of Policy and Resource Management, *Can We Delay a Greenhouse Warming?* September 1983. Written by Stephen Seidel and Dale Keyes.

36. World Meteorological Organization, *Report of the International Conference on the Assessment of the Role of Carbon Dioxide.* Villach, Austria, Oct 9–15, 1985, WMO Doc. No. 661 (1986).

37. Daniel Bodansky, "Bonn Voyage: Kyoto's Uncertain Revival," *National Interest* 65 (2001): 48.

38. World Conference on the Changing Atmosphere, Proceedings, Toronto, June 27–30, 1988 WMO Doc. 710 (1989).

39. Bodansky, "Bonn Voyage," p. 51.

40. Senate Committee on Energy and Natural Resources, "Statement of James Hansen," *Greenhouse Effect and Global Climate Change,* June 23, 1988, 100 Cong. 2nd Session (1988).

41. Michael Weisskopf, "Baker Sitting Out Global Warming Debate," *Washington Post,* November 9, 1990, A25.

42. Philip Shabecoff, "White House Admits Censoring Testimony," *New York Times,* May 9, 1989, C1.

43. George Bush, "Remarks to the Intergovernmental Panel on Climate Change," February 5, 1990, *Public Papers of the Presidents of the United States, George Bush 1990* (Office of the Federal Register, National Archives and Records Administration), Vol. 1, pp. 159, 158.

44. Michael Weisskopf, "Shift on Warming Sought," *Washington Post,* February 3, 1990, A1.

45. "White House Denies Dissension," *Facts on File,* February 9, 1990, p. 81.

46. Bob Davis, "Bid to Slow Global Warming Could Cost U.S. $200 Billion a Year," *Wall Street Journal,* April 16, 1990, B4.

47. Michael Weisskopf, "Climate Meeting Ends in Controversy," *Washington Post,* April 19, 1990, A19.

48. Herb Brody, "The Political Pleasures of Engineering: an Interview with John Sununu," *Technology Review,* August-September 1992, pp. 22–28.

49. Rose Gutfeld, "Earth Summitry," *Wall Street Journal,* May 27, 1992, A1; Lawrence E. Susskind, Environmental Diplomacy (New York: Oxford, 1994), p. 40.

50. Lynne M. Jurgielewicz, *Global Environmental Change and International Law* (Lanham, MD: University Press of America, 1996), p. 35; Gary C. Bryner, *From Promises to Performance* (New York: W. W. Norton, 1997), p. 20.

51. Lucia Mouat, "U.S. Wins Concessions on Global Warming, but Loses Goodwill," *Christian Science Monitor,* May 4, 1992, p. 9.

52. "Key Remarks and Speeches from the Rio Earth Summit," *Facts on File,* June 18, 1992, p. 442.

53. Keith Schneider, "Gore Meets Resistance in Effort for Steps on Global Warming," *New York Times,* April 19, 1993, A17.

54. Kevin Trenberth, "The Use and Abuse of Climate Models," *Nature,* March 13, 1997, pp. 131–133.

55. Alexis Simendinger, "In the Global-warming Box," *National Journal,* June 16, 2001.

56. Timothy O'Riordan, "The Early Bush Presidency and Climate Change Politics," *Environment,* June 2001.

57. George W. Bush, "Letter to Senators Hagel, Helms, Craig and Roberts," Press Release, White House, March 13, 2001.

58. Robert J. Samuelson, "The Kyoto Delusion," *Washington Post,* June 21, 2001, p. A 25.

59. Ibid.

60. O'Riordan, "The Early Bush Presidency."

61. Charlotte Schubert, "Global Warming Debate Gets Hotter," *Science News,* June 16, 2001.

62. Andrew Revkin, "Report by the EPA Leaves Out Data on Climate Change," New York Times, June 19, 2003; "Emissions Omissions," Boston Globe, June 21, 2003.

63. *EPA's Draft Report on the Environment 2003: Technical Document* www.epa. gov/indicators/ roe/index.htm, June 30, 2003, p. 1–38.

64. Gwynne Dyer, "Global Accord on Emissions Still a Crisis Away," *The Toledo Blade,* December 6, 2005.

65. Peter Baker, "Russia Backs Kyoto to Get on Path to Join WTO," *Washington Post,* May 22, 2004, A15.

66. "Get Real on Climate Change," *The Observer,* October 30, 2005.

67. David Appell, "Behind the Hockey Stick," *Scientific American,* February 21, 2005.

68. Andrew C. Revkin, "Politics Reasserts Itself in the Debate over Climate Change," *New York Times,* August 5, 2003; J. R. Pegg, "GOP Senators Blame Nature for Climate Change," *Environment News Service,* July 30, 2003, www. oneworld.net/article/view/64634/1/.

69. *Wall Street Journal,* December 4, 1997.

70. William Stevens, "Skeptic Asks: Is It Really Warmer?" *New York Times* December 1, 1998.

71. Robert Cowen, "Scientists Can't Quite Finger Humans as Cause of Earth's Rising Temps," *Christian Science Monitor,* December 22, 1997.

72. Reuters, "Scientists See Sun, Not People, Behind Global Warming," CNN, May 5, 1998, http://cnn.com/tech/science/.

73. William Stevens, "Study of Ocean Currents Offers Clues to Global Climate Shifts," *New York Times,* March 18, 1997, C 1.

74. P. J. Michaels and P. C. Knappenberger, "Human Effect on Global Climate?" *Nature* 384 (1996): 522–23; P. J. Michaels, "The Consequences of Kyoto," *Cato Policy Analysis* No. 307 (May 7, 1998).

75. Michaels, "The Consequences of Kyoto," pp. 8, 11.

76. Martha M. Hamilton, "Global Warming Gets a Second Look," *Washington Post,* March 3, 1998.

— 7 —

Conclusion

The prophetic tradition continues. Public opinion polls show that a large portion of Americans believe in the biblical version of the Apocalypse. To some like Ronald Reagan, belief is literal and sincere. George W. Bush claims a conventional view of Christianity and has discreetly refrained from the comments on the end of the world that brought criticism to Reagan. Jimmy Carter was another believer who also avoided explicit references to the Apocalypse but was in tune with its prophecies of doom and calls for moral renewal.

Environmentalists, who do not often advertise their belief in the Bible, certainly not literally, nevertheless share the prophetic tradition. They often proclaim the most dire warnings of how no birds will sing, how the globe will warm up so life will suffer and the sea will rise, how overpopulation will cause starvation, and how bombs will kill hundreds of millions directly and cause nuclear winter. Many of the predictions are of war: Shortages of energy and food will lead to combat; countries with excess people will attack their neighbors for living space. Although the predictions may appear overstated, one can hardly doubt their sincerity.

Modern environmental apocalypses, of course, are not the same as the biblical ones. They are based on science instead of religion. Their authors are not ancient men of God, but physicists, statisticians, and econometricians. The ordinary people have not erred by straying from God's way, but by polluting the earth, having too many babies, building atom bombs, or burning too much fossil fuel. The modern apocalypses disclose information about the future that has been discovered by scientific research, as opposed to being a secret known only to God's prophets. Both religious and scientific ones predict catastrophe and have a sense of urgency. While both call for moral renewal, the modern prophets also call for specific practical steps to be taken. The authors of the modern versions do not write them as allegory, but to be taken literally.

Modern environmental apocalypses have been forecast both by individual prophets and by government bodies. Aurelio Peccei formulated the concept of the Predicament of Mankind single-handedly. M. King Hubbert forecast the shortage of oil largely by himself. Paul Ehrlich coined the term population bomb. The five authors of the TTAPS report laid out the threat of nuclear winter. Svante Arrhenius discovered global warming and its cause of burning coal. Other times commissions have played the role of prophet. The National Conservation Commission in 1909 and the Paley Commission in 1952 predicted the energy crisis years ahead. The Draper Committee forecast the population explosion. Virtually all these modern threats have depended on scientific and statistical analysis.

To various degrees the American government has tried to develop plans, often in cooperation with other countries. The success of the plans varies and often disappoints. The Nixon-Ford Project Energy Independence was incomplete, inconsistent, ignored Europe and Japan, and was not based on sound economics. Carter's elaborate national energy plan was neither comprehensive nor consistent. Neither plan proved successful, and they were promptly abandoned. As a counter example, in passing the 1924 Johnson Reed Act, Congress established a coherent immigration policy with a clear goal that worked for 20 years or more and was replaced because the goal of maintaining an ethnic balance fell out of favor. In formulating solutions, a previous model may be influential. The clearest example is the ozone hole for greenhouse gases. In the cases of energy planning, the previous model was wartime planning with a controlled economy rather than a free market. The Club of Rome was also enamored of control from the top, in this case owing a debt to Italian Fascism.

With the tendency of energy plans and the Club of Rome report to imitate a military model, it is ironic that the military itself did not plan cohesively for atomic warfare. After Hiroshima the manufacture of weapons went forward piecemeal, until in 1954 Eisenhower and Dulles realized that 2,000 were in the arsenal. At this time, they slapped on the label of massive retaliation and combined it with brinkmanship. As demonstrated in Korea,

Cuba, and Vietnam, it was not effective for limited war. While some argue that the fear of retaliation kept the superpowers from attacking each other, little credit can go to the planning aspect of this balance of terror. During the long Cold War, the strict military secrecy was a major cause of the lack of comprehensive planning. On the other side, the attempts to disarm were not very well planned either. Here most of the blame goes to the Soviet Union, because for much of the time its leadership was too unstable or too old to address the problem.

International cooperation is a necessity for dealing with apocalyptic problems throughout their life cycles. The first stage of the scientific discovery and testing of the evidence often transpires in international scientific societies. Once past this stage and on to the policy aspects, in recent years the life cycle has moved on to a technical workshop, often sponsored by a UN organization like the World Meteorological Organization, the UN Environmental Program, or the UN Population Fund. At this stage the parties often form a special group like the IPCC. Next comes a big conference with diplomats and officials in addition to the scientists. Nongovernmental groups attend, too. Some conferences have become quite big and theatrical, like the Earth Summit in Rio and the global warming conference in Kyoto. Indeed a criticism of recent ones like the Earth Summit in Johannesburg is that they have become so large that they can not accomplish much of substance.

The workshops, technical meetings, and larger conferences may go on for years, depending on the evidence, the degree of scientific consensus, and the willingness of the different countries to reach agreement. The culmination is a framework treaty with many countries at one of the large conferences, such as the Framework Convention on Climate Change signed at Rio. The typical framework convention provides for a Conference of the Parties, which meets annually to coordinate, review research, and prepare the more specific protocols. For climate change, it took the Conference of the Parties six years to write the Kyoto Protocol. The treaties, conventions, and protocols have various degrees of specificity, since ambiguity can allow compromise. The basic elements of the Kyoto Protocol came from the Toronto meeting nine years before and are subject to differing interpretations about the starting point, the credit a country deserves for nuclear energy or forests, and joint implementation. Ozone depletion is another case that went through the full cycle. Although there have been many big conferences on population, the problem is not so amenable to a treaty. The Nuclear Non-Proliferation Treaty of 1968 fits the typical pattern. But in the case of disarmament, only two countries were important, making the key to bargaining the agreement of the United States and the Soviet Union.

Looking back, some of the problems are under control from a practical perspective, and some are not. Control of ozone depletion appears to be a success. The Nuclear Non-Proliferation and disarmament treaties appear largely to be successes. Population control is not a success measured in terms of formal agreements and institutions, but the rate of population growth is declining anyway. It is too early to determine success to prevent global warming. The SST problem went away on its own. Energy is not an acute problem today, but little credit should go to the formal attempts to deal with it. Most of the aspects of the Predicament of Mankind that challenged Aurelio Peccei remain, but their severity is less.

The next factor is the ease and cheapness of the solution. Eliminating CFCs that destroyed the ozone layer was not expensive. Their consumption was limited to air conditioning, refrigerators, and spray cans. The chemical companies soon realized that the cost would not be high, and with more research, the substitute chemicals proved to be cheaper. This is the opposite with the problem of global warming, which will be extremely expensive. Carbon is burned in virtually every industrial process as well as for automobiles and home heating. Every part of the economy is affected. Unlike CFCs, carbon and other greenhouse gases are so pervasive that their use can only be reduced, not eliminated. Overpopulation is a problem that is not easy, because it depends on social custom, education, and prosperity as well as its technical equipment. Simply paying for birth control pills or contraception for India alone would cost billions of dollars. Nuclear disarmament presents a happier situation, since having fewer weapons is cheaper than having more. The impediment is the fear that reducing the weapons would leave the United States militarily vulnerable. Russia faces the same difficulty from its side.

When the United States has clear goals, success is easier. Again, phasing out CFCs to protect the ozone layer is exemplary. The goal was the rapid elimination of these chemicals, and few other aspects mattered. American goals to control the greenhouse effect are uncertain. The greenhouse gas target is poorly understood, the starting and finishing years are debated, and the interaction between carbon dioxide and the other greenhouse gases like methane and nitrous oxide is unknown. The United States does not know how it will limit burning coal, oil, and gas. While it seems intrigued with the idea of tradable carbon units as a panacea, this scheme is still hypothetical and fraught with difficulties. American goals for food and overpopulation in the third world are positive but vague. People should not starve or have too many babies, but the details are not settled.

American autonomy to solve apocalyptic problems by itself varies. Because it produced half the CFCs in the world, the American phaseout had a big impact. The United States produces 20–25 percent of the carbon dioxide, which is only half as much as the case for ozone protection. American ability to control population in India, China, and Africa is small. With respect to energy, the United States imports 60 percent of its oil, greatly limiting its ability to act autonomously. Because the United States has the biggest economy in the world, it can have power in a negative sense. As a practical matter, the Kyoto Protocol cannot succeed without its participation, much as the Europeans dislike the fact.

Because these are apocalyptic threats affecting the entire world, cooperation with other countries is crucial. Often the United States finds natural commonality with other industrial democracies like Britain, France, Germany, and Japan. But goals are not shared 100 percent. It took a while for Britain, France, and Italy to join the phaseout of CFCs. At present the Europeans are quite angered by the American reluctance to sign and implement the Kyoto Protocol. For 40 years the United States and the Soviet Union shared a general goal of reducing nuclear weapons, but how to do so without losing a military advantage stymied progress. The United States sometimes may share goals with the third world. China agrees that overpopulation is a threat and has instituted strict controls. India agrees it is a threat, but has been much less effective. On other occasions, goals may be opposed. With respect to the greenhouse effect, China and India have been very determined to avoid limits to the carbon dioxide they can emit.

Looking back, it is easy to see that many environmental catastrophes have been ignored in spite of multiple warnings, and that the results have been bad. Planning to prevent them would have been good, but often does not occur.

Selected Bibliography

Barney, Gerald O. (study director). *The Global 2000 Report to the President*. Washington, DC: Council on Environmental Quality and the Department of State, 1980.

Benedick, Richard Elliot. *Ozone Diplomacy*. Cambridge, MA: Harvard University Press, 1991.

Brundtland, Gro Harlem (chair). *Our Common Future* Report of the World Commission on Environment and Development. New York: Oxford University Press, 1987.

Ehrlich, Paul R. *The Population Bomb*. New York: Ballantine, 1970.

Ehrlich, Paul R., and Anne H. Ehrlich. *The Population Explosion*. New York: Simon and Schuster, 1990.

Fitzgerald, Frances. *Way Out There in the Blue: Reagan, Star Wars and the End of the Cold War*. New York: Simon and Schuster, 2000.

Hatfield, Craig. "Oil: Back on the Global Agenda." *Nature,* May 8, 1997.

Horwitch, Mel. *Clipped Wings: the American SST Conflict*. Cambridge, MA: MIT Press, 1982.

Kahn, Herman. *On Thermonuclear War*. Princeton, NJ: Princeton University Press, 1960.

Kahn, Herman, and Anthony J. Wiener. *The Year 2000: A Framework for Speculation on the Next Thirty-Three Years*. New York: Macmillan, 1967.

Levy, Elizabeth. *The SST Story*. New York: Delacorte Press, 1973.

Livi-Bacci, Massimo. *A Concise History of World Population* (2nd ed.). London: Blackwell, 1997.

Lomborg, Bjorn. *The Skeptical Environmentalist*. New York: Cambridge University Press, 2001.

Mazur, Laurie Ann, ed. *Beyond the Numbers*. Washington, DC: Island Press, 1994.

Meadows, Donella H., Dennis L. Meadows, Jorgen Randers, and William W. Behrens. *The Limits to Growth*. New York: Universe Books, 1972.

Mesarovic, Mihajlo, and Eduard Pestel. *Mankind at the Turning Point*. New York, Dutton, 1974.

Meyer, David S. *A Winter of Discontent: The Nuclear Freeze and American Politics*. New York: Praeger, 1990.

Michaels, Patrick J. *Meltdown: The Predictable Distortion of Global Warming by Scientists, Politicians, and the Media*. Washington, DC: Cato Institute, 2004.

Norris, Robert S., and Hans M. Kristensen. "Nuclear Notebook: Global Nuclear Stockpiles, 1945–2006." *Bulletin of the Atomic Scientists* 62 (2006): 64–66.

Paul, Diane B. *Controlling Human Heredity*. Atlantic Highlands, NJ: Humanities Press, 1995.

Piotrow, Phyllis Tilson. *World Population Crisis*. New York: Praeger, 1973.

Ponte, Lowell. *The Cooling*. Englewood Cliffs, NJ: Prentice Hall, 1976.

Revkin, Andrew. *Global Warming*. New York: Abbeville Press, 1992.

Sagan, Carl, and Richard Turco. *A Path Where No Man Thought: Nuclear Winter and the End of the Arms Race*. New York: Random House, 1990.

Simon, Julian L. *Hoodwinking the Nation*. New Brunswick: Transaction, 1999.

———. *Population Matters*. New Brunswick, NJ: Transaction, 1990.

Solinger, Rickie, ed. *Abortion Wars*. Berkeley: University of California Press, 1998.

Staggenborg, Suzanne. *The Pro-Choice Movement*. New York: Oxford University Press, 1991.

Wilson, Carroll L. (project director). *Energy: Global Prospects, 1985-2000: Report of the Workshop on Alternative Energy Strategies (WAES)*. New York: McGraw-Hill, 1977.

Index

About the Author

DAVID HOWARD DAVIS is Professor of Political Science at the University of Toledo. He formerly taught at the University of Wyoming, Cornell, and Rutgers. He has been an energy consultant and has served in the U.S. Department of the Interior. His experience includes stints as an analyst at the Congressional Research Service and as a faculty fellow at the General Accounting Office. Davis is the author of three previous books.